Further Praise for

THE BONOBO AND THE ATHEIST

"In this richly observed and intelligent book, de Waal ponders our natural receptiveness to religion, how religion evolved and what if anything might take its place." —*Scientific American*

"Compelling. . . . Blows the idea of top-down morality out of the water." —*Independent* (UK)

"De Waal isn't telling religious people that they're wrong to love their fellow humans (and non-humans) because God tells them to. He's just explaining why the rest of us feel the same way, whether God speaks to us or not." —*Psychology Today*

"A primatologist who has spent his career studying chimpanzees and bonobos, two of humanity's closest living relatives, Mr de Waal draws on a lifetime of empirical research. His data provides plenty of evidence that religion is not necessary in order for animals to display something that looks strikingly like human morality." —*Economist*

"De Waal states his argument for the evolution of human empathy with the sophistication of a well-grounded, risk-taking scientist who can venture into philosophy."
—*Nature*

"Dr. de Waal resists hard-line atheism and religion alike. . . . Refreshingly undogmatic."
—*New York Times*

"An elegant stylist, de Waal writes about these animals with great affection and respect."
—*Boston Globe*

"Frans de Waal's new book carries the important message that human kindness is a biological feature of our species and not something that has to be imposed on us by religious teaching."
—Desmond Morris, author of *The Naked Ape*

ALSO BY FRANS DE WAAL

The Age of Empathy (2009)

Primates and Philosophers (2006)

Our Inner Ape (2005)

My Family Album (2003)

The Ape and the Sushi Master (2001)

Bonobo (1997)

Good Natured (1996)

Peacemaking among Primates (1989)

Chimpanzee Politics (1982)

THE BONOBO
and
THE ATHEIST

In Search of Humanism Among the Primates

FRANS DE WAAL

with drawings by the author

W. W. NORTON & COMPANY • NEW YORK LONDON

Printed in the United States of America
First published as a Norton paperback 2014

Reproductions of Hieronymus Bosch's paintings were derived from freely licensed file repositories in the public domain, such as Wikimedia Commons, using fragments of larger paintings: *The Garden of Earthly Delights*, *Ascent of the Blessed*, and *The Last Judgment*. The cartoon of T. H. Huxley by Carlo Pellegrini first appeared in *Vanity Fair* of 28 January 1871. All other illustrations are pencil and ink drawings or photographs made by the author, except for two photographs of bonobo "Vic," which are reproduced with permission of the photographer, Marian Brickner.

For information about special discounts for bulk purchases, please contact W. W. Norton Special Sales at specialsales@wwnorton.com or 800-233-4830

Manufacturing by RR Donnelley, Harrisonburg, VA
Book design by Dana Sloan
Production manager: Anna Oler

Library of Congress Cataloging-in-Publication Data

Waal, F. B. M. de (Frans B. M.), 1948–
The bonobo and the atheist : in search of humanism among the primates / Frans de Waal ; with drawings by the author. — First edition.
pages cm
ISBN 978-0-393-07377-5 (hardcover)
1. Bonobo—Behavior. 2. Altruistic behavior in animals. 3. Ethics. 4. Empathy. 5. Human-animal relationships. 6. Waal, F. B. M. de (Frans B. M.), 1948– 7. Primatologists—Biography. I. Title.
QL737.P96W315 2013
205'.693—dc23
2012048868

ISBN 978-0-393-34779-1 pbk.

W. W. Norton & Company, Inc.
500 Fifth Avenue, New York, N.Y. 10110
www.wwnorton.com

W. W. Norton & Company Ltd.
Castle House, 75/76 Wells Street, London W1T 3QT

1 2 3 4 5 6 7 8 9 0

For Catherine, my favorite primate

CONTENTS

THE BONOBO AND THE ATHEIST

Chapter 1

EARTHLY DELIGHTS

Is man only a blunder of God? Or is God only a blunder of man?

—Friedrich Nietzsche[1]

I was born in Den Bosch, the Dutch city after which Hieronymus Bosch named himself.[2] This doesn't make me an expert on the painter, but having grown up with his statue on the market square, I have always been fond of his surrealist imagery, his symbolism, and how it relates to humanity's place in the universe under a waning influence of God.

His famous triptych in which naked figures frolic around, *The Garden of Earthly Delights*, is a tribute to paradisiacal innocence. The middle tableau is far too happy and relaxed to fit the interpretation of depravity and sin advanced by puritan experts. It shows humanity free from guilt and shame either before the Fall or without any Fall at all. For a primatologist like myself, the nudity, the allusions to sex and fertility, the plentiful birds and fruits, and the moving about in groups are thoroughly familiar, and hardly in need of a religious or moral interpretation. Bosch seems to have depicted us in our natural state, while reserving his moralistic outlook for the right-hand panel, in which he punishes *not* the frolickers from the middle panel but monks, nuns, gluttons, gamblers, warriors, and drunkards. Bosch was no fan of the clergy and their avarice, which explains a small detail in which a man resists signing his fortune away to a pig veiled like a Dominican nun. The poor figure is said to be the painter himself.

Five centuries later, we remain embroiled in debates about the place of religion in society. As in Bosch's day, the central theme is morality. Can we envision a world without God? Would this world be good? Don't think for one moment that the current battle lines between fundamentalist Christianity and science are determined by evidence. One has to be pretty immune to data to doubt evolution, which is why books and documentaries aimed at convincing the skeptics are a waste of effort. They are helpful for those prepared to listen, but fail to reach their target audience. The debate is less about the truth than about how to handle it. For those who believe that morality comes straight from God the creator, acceptance of evolution would open a moral abyss. Listen to the Reverend Al Sharpton debating the late atheist firebrand Christopher Hitchens: "If there is no order to the universe, and therefore some being, some force that ordered it, then who determines what is right or wrong? There is nothing immoral if there's nothing in charge."[3] Similarly, I have heard people echo Dostoevsky's Ivan Karamazov, exclaiming, "If there is no God, I am free to rape my neighbor!"

Perhaps it's just me, but I am wary of any persons whose belief system is the only thing standing between them and repulsive behavior. Why not assume that our humanity, including the self-control needed for a livable society, is built into us? Does anyone truly believe that our ancestors lacked social norms before they had religion? Did they never assist others in need, or complain about an unfair deal? Humans must

In the lower right-hand corner of *The Garden*, Bosch depicted himself resisting a pig dressed like a nun, who tries to seduce him with kisses. She is offering salvation in return for his estate (hence the pen, ink, and official-looking paper). *The Garden* was painted around 1504, about a decade before Martin Luther galvanized protest against such church practices.

have worried about the functioning of their communities well before current religions arose, which occurred only a couple of millennia ago. Biologists are unimpressed by that kind of timescale.

The Dalai Lama's Turtle

The above introduced a blog entitled *Morals without God?* on the *New York Times'* website, in which I argued that morality antedates religion and that much can be learned about its origin by considering our fellow primates.[4] Contrary to the customary blood-soaked view of nature, animals are not devoid of tendencies that we morally approve of, which to me suggests that morality is not as much of a human innovation as we like to think.

This being the topic of the present book, let me lay out its themes by describing the week that followed my blog's publication, including a trip to Europe. Right before this, however, I attended a meeting between science and religion at Emory University, in Atlanta, where I work. The occasion was a forum with the Dalai Lama on his favorite theme: compassion. Being compassionate seems to me an excellent recommendation for life; hence I welcomed the message of our honorable guest. As the first discussant, I was seated next to him surrounded by a sea of red and yellow chrysanthemums. I had been instructed to address him as "your holiness," but to speak of him to others as "his holiness," which I found sufficiently confusing that I avoided all forms of address. One of the most admired men on the planet dropped his shoes and folded his legs under him in his chair, put on a huge baseball cap color-matched to his orange robe, while an audience of over three thousand people hung on his every word. Before my presentation, I had been appropriately deflated by the organizers' reminding me that no one had come to hear *me* speak, and that all those people were there only for *his* pearls of wisdom.

In my remarks, I reviewed the latest evidence for animal altruism. For example, apes will voluntarily open a door to offer a companion access to food, even if they lose part of it in the process. And capuchin monkeys are prepared to seek rewards for others, as we see when we place two of them side by side, while one of them barters with us with differently colored tokens. One token rewards only the monkey itself, whereas the other rewards both monkeys. Soon, the monkeys prefer the "prosocial" token. This is not out of fear, because dominant monkeys (who have least to fear) are in fact the most generous.

Good deeds also occur spontaneously. An old female, Peony, spends her days outdoors with other chimpanzees at the Yerkes Primate Center's field station. On bad days, when her arthritis is flaring up, she has trouble walking and climbing, but other females help her out. Peony may be huffing and puffing to get up into the climbing frame in which several apes have gathered for a grooming session. But an unrelated younger female moves behind her, placing both hands on her ample behind to push her up with quite a bit of effort, until Peony has joined the rest.

We have also seen Peony get up and slowly move toward the water spigot, which is at quite a distance. Younger females sometimes run ahead of her, take in some water, then return to Peony and give it to her. At first, we had no idea what was going on, since all we saw was one female placing her mouth close to Peony's, but after a while the pattern became clear: Peony would open her mouth wide, and the younger female would spit a jet of water into it.

Such observations fit the emerging field of animal empathy, which deals not only with primates but also with canines, elephants, and even rodents. A typical example is how chimpanzees console distressed parties, hugging and kissing them, which is so predictable that we have documented literally thousands of cases. Mammals are sensitive to each other's emotions and react to those in need. The whole reason people fill their homes with furry carnivores, and not with, say, iguanas and turtles, is that mammals offer something no reptile ever will. They give affection, they want affection, and they respond to our emotions the way we do to theirs.

Up to this point, the Dalai Lama had listened attentively, but now he lifted his cap to interrupt me. He wanted to hear more about turtles. These animals are a favorite of his, because they supposedly carry the world on their backs. The Buddhist leader wondered whether turtles, too, know empathy. He described how the female sea turtle crawls onto land to look for the best spot to lay her eggs, thus showing concern for future young. How would the mother behave if she ever encountered her offspring? the Dalai Lama wondered. To me, the process suggests that turtles have been preprogrammed to seek out the best environment for incubation. The turtle digs a hole in the sand above the tide line, deposits her eggs and covers them, packing the sand tight with her rear flippers, and then leaves the nest behind. The hatchlings emerge a few months later to rush to the ocean under the moonlight. They never get to know their mother.

Empathy requires awareness of the other and sensitivity to the other's needs. It probably started with parental care, like that found in the mammals, but there is also evidence for bird empathy. I once visited the Konrad Lorenz Research Station, in Grünau, Austria, which keeps ravens in large aviaries. These are impressive birds, especially when they sit on your shoulder with their powerful black beak right next to your face! It brought back memories of the tame jackdaws I had kept as a student: much smaller birds from the same corvid (crow) family. In Grünau, scientists follow spontaneous fights among the ravens and have seen bystanders respond to distress. Losers can count on some cozy preening or beak-to-beak nudging from their friends. At the same station, free-ranging descendants of Lorenz's flock of geese have been equipped with transmitters to measure their heart rate. Since every adult goose has a mate, that offers a window on empathy. If one bird confronts another in a fight, its partner's heart starts racing. Even if the partner is in no way involved, its heart betrays concern about the quarrel. Birds, too, feel each other's pain.

If both birds and mammals have some measure of empathy, that capacity probably goes back to their reptilian ancestors. Not just any reptiles, though, because most lack parental care. One of the surest signs of a caring attitude, according to Paul MacLean, the American neuroscientist who named the limbic system the seat of the emotions, is the "lost call" of young animals. Young monkeys do it all the time:

left behind by mom, they call until she returns. They look miserable, sitting all alone on a tree limb, giving a long string of plaintive "coo" calls with pouted lips directed at no one in particular. MacLean noted the absence of the "lost call" in most reptiles, such as snakes, lizards, and turtles.

In a few reptiles, however, the young do call when upset or in danger, so that mom will take care of them. Have you ever held a baby alligator? Be careful, because they have a good set of teeth, but they also utter throaty barks when upset, which may bring the cow (mother) flying out of the water. That will teach you to doubt reptilian feelings!

I mentioned this to the Dalai Lama, saying that we expect empathy only in animals with attachments, and that few reptiles qualify. I am not sure this satisfied him, because of course he wanted to know about turtles, which look so much cuter than those ferocious toothy monsters of the Crocodilian family. Appearances are deceptive, though. Some members of this family gently transport their young in their big jaws or on their backs and defend them against danger. They sometimes even let them snatch pieces of meat from their mouth. The dinosaurs, too, cared for their young, and plesiosaurs—giant marine reptiles—may even have been viviparous, giving birth to a single live offspring in the water, as whales do today. From everything we know, the smaller the number of offspring an animal produces, the better it will take care of them, which is why plesiosaurs are thought to have been doting parents. So, by the way, are birds, which science regards as feathered dinosaurs.

Pressing me even further, the Dalai Lama jumped to butterflies and asked about their empathy, upon which I couldn't resist joking, "They don't have time, they live just one day!" The short life of butterflies is actually a myth, but whatever these insects feel about each other, I doubt it has much to do with empathy. This is not to minimize the larger thrust behind the Dalai Lama's question, which was that all ani-

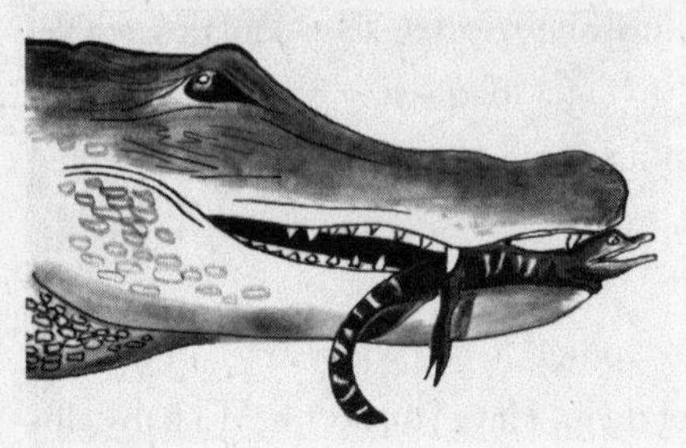

Few reptiles have parental care, but the crocodile family does. A female alligator safely transports one of her young.

mals do what is best for themselves and their offspring. In this sense, all life is caring, perhaps not consciously caring, but caring nonetheless. He was getting at the idea that compassion goes to the root of what life is all about.

Greeting Mama

After this, the forum moved on to other topics, such as how to measure compassion in the brains of Buddhist monks who have meditated on it all their lives. Richard Davidson from the University of Wisconsin related how monks straight from Tibet balked at his invitation to submit to neuroscience since, clearly, compassion didn't take place in the brain but in the heart! Everyone felt this was hilarious, and the monks in the audience shrieked with laughter. But the monks had a point. Davidson subsequently discovered the connection between mind and heart: compassion meditation brings about a quicker heart rate upon hearing sounds of human suffering.

I had to think of the geese. But I also sat there wondering at this auspicious meeting of minds. In 2005, the Dalai Lama himself had spoken about the need to integrate science and religion, telling thousands of scientists at the annual meeting of the Society for Neuroscience, in Washington, how much trouble society has in keeping up with their groundbreaking research: "It is all too evident that our moral thinking simply has not been able to keep pace with such rapid progress in our acquisition of knowledge and power."[5] What a refreshing departure from attempts to drive a wedge between religion and science!

This topic was on my mind as I prepared for Europe. I had barely received a blessing and a *khata* (a long white silk scarf) around my neck, and seen the Dalai Lama off in his limousine with heavily armed guards, and I was on my way to Ghent, a beautiful old city in the Flemish part of Belgium. This region is culturally closer to the southern part of the Netherlands, where I am from, than the part to the north that we call Holland. All of us speak the same language, but Holland is Calvinist, whereas the southern provinces were kept Catholic in the sixteenth century by the Spanish, who brought us the Duke of Alva and the Inquisition. Not the silly "Nobody expects the Spanish Inquisition!" of *Monty Python*, but one that would put actual thumbscrews on you if you so much as doubted Mary's virginity. Not allowed to draw blood, the inquisitors loved the *strappado*, or reverse hanging, in which a victim is hung by wrists tied behind his back and a weight is attached to his ankles. This treatment is sufficiently debilitating that one soon abandons any preconceived notions about the link between sex and conception. Lately, the Vatican has been on a campaign to soften the Inquisition's image—they did not kill *every* heretic, they followed Standard Operating Procedures—but the Jesuits in charge surely could have used some compassion training.

This ancient history also explains, by the way, why one will look in vain for Bosch paintings in the lowlands. Most hang in the Prado, in Madrid. It is thought that the Iron Duke obtained *The Garden* when, in 1568, he declared the Prince of Orange an outlaw and confiscated all of his properties. The duke then left the masterpiece to his son, from whom it went to the Spanish state. The Spanish adore the painter they call *El Bosco*, whose imagery inspired Joan Miró and Salvador Dalí. On my first visit to the Prado, I could not really enjoy Bosch's work, since all I could think was "Colonial plunder!" To its credit, the museum has now digitized the popular painting at an incredibly high resolution so that everyone can "own" it through Google Earth.

After my lecture in Ghent, fellow scientists took me on an impromptu visit to the world's oldest zoo collection of bonobos, which started at Antwerp Zoo and is now located in the animal park of Planckendael. Given that bonobos are native to a former Belgian colony, their presence in Planckendael is hardly surprising. Bringing specimens from Africa, dead or alive, was another kind of colonial plunder, but without it we might never have learned of this rare ape. The discovery took place in 1929, in a museum not far from here, when a German anatomist dusted off a small round skull labeled as that of a young chimp, which he recognized as an adult with an unusually small head. He quickly announced a new subspecies. Soon his claim was overshadowed, however, by the even more momentous pronouncement by an American anatomist that we had an entirely new species on our hands, one with a strikingly humanlike anatomy. Bonobos are more gracefully built and have longer legs than any other ape. The species was put in the same genus, *Pan*, as the chimpanzee. For the rest of their long lives, both scientists illustrated the power of academic rivalry by never agreeing on who had made this historic discovery. I was in the room when the American stood up in the midst of a symposium on bonobos to declare, in a voice quavering with indignation, that he had been "scooped" half a century before.

The German scientist had written in German and the American in English, so guess whose story is most widely cited? Many languages feel the pinch of the rise of English, but I was happily chatting in Dutch, which despite decades abroad still crosses my lips a fraction of a second faster than any other language. While a young bonobo swung on a rope in and out of view, getting our attention by hitting the glass each time he passed, we commented on how much his facial expression resembled human laughter. He was having fun, especially if we jumped back from the window, acting scared. We now find it impossible to imagine that the two *Pan* species were once mixed up. There is a famous photograph of the American expert Robert Yerkes, with two young apes on his lap, both of whom he considered chimps. This was before the bonobo was known. Yerkes did remark how one of those two apes was far more sensitive and empathic than any other he knew, and perhaps also smarter. Calling him an "anthropoid genius," he wrote his book *Almost Human* largely about this "chimpanzee," not knowing that he was in fact dealing with one of the first live bonobos to have reached the West.

The Planckendael colony shows the difference with chimpanzees right away, because it is led by a female. The biologist Jeroen Stevens told me how the atmosphere in the group had turned more relaxed since their longtime alpha female, who had been a real iron lady, had been sent off to another zoo. She had terrified most other bonobos, especially the males. The new alpha has a nicer character. The exchange of females between zoos is a new and commendable trend that fits the natural bonobo pattern. In the wild, sons stay with their mothers through adulthood, whereas daughters migrate to other places.

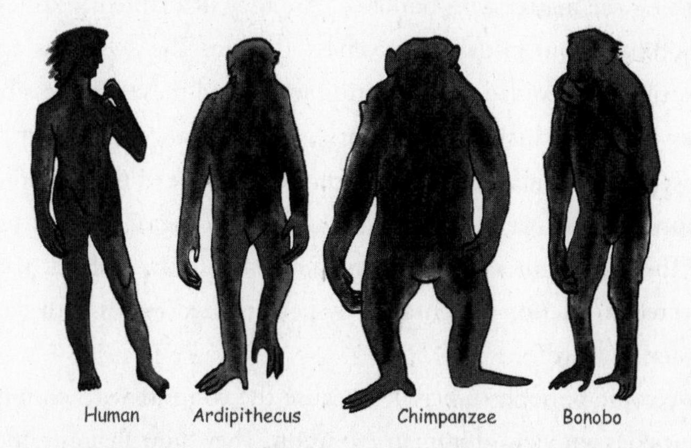

During human evolution, bipedal locomotion demanded longer legs. Of all the apes, the arm-to-leg ratio of bonobos most resembles that of our ancestor *Ardipithecus* (drawing not to scale: modern humans are taller than the rest).

For years, zoos had been moving males around, thus causing disaster upon disaster, because male bonobos get hammered in the absence of their mom. Those poor males often ended up in isolation in an off-display area of zoos in order to protect their lives. A lot of problems are being avoided by keeping males with their mothers and respecting their bond.

This goes to show that bonobos are no angels of peace. But it also indicates how much the males are "mama's boys," something not everyone approves of. Some men feel affronted by matriarchal apes with "wimpy" males. After a lecture in Germany, a famous old professor in my audience barked, "*Was ist vrong* with those males?!" It is the fate of the bonobo to have burst on the scientific scene at a time when anthropologists and biologists were busy emphasizing violence and warfare, hence scarcely interested in peaceful primate kin. Since no one knew what to do with them, bonobos quickly became the black sheep of the human evolutionary literature. An American anthropologist went so far as to recommend that we simply ignore them, given that they are close to extinction anyway.[6]

Holding a species' imminent demise against it is extraordinary. Is something the matter with bonobos? Are they ill adapted? Extinction says nothing about initial adaptiveness, though. The dodo was doing fine until sailors landed on Mauritius and found these flightless birds an easy (if repugnant) meal. Similarly, all of our ancestors must have been well adapted at some point, even though none of them is around anymore. Should we stop paying attention to them? But we never stop. The media go crazy each time a minuscule trace of our past is discovered, a reaction encouraged by personalized fossils with names like Lucy and Ardi.

I welcome bonobos precisely because the contrast with chimpanzees enriches our view of human evolution. They show that our lineage is marked not just by male dominance and xenophobia but also by a love of harmony and sensitivity to others. Since evolution occurs through both the male and the female lineage, there is no reason to measure human progress purely by how many battles our men have won against other hominins.[7] Attention to the female side of the story would not hurt, nor would attention to sex. For all we know, we did not conquer other groups, but bred them out of existence through love rather than war. Modern humans carry Neanderthal DNA, and I wouldn't be surprised if we carry other hominin genes as well. Viewed in this light, the bonobo way doesn't seem so alien.

Leaving those gentle apes behind, I next stopped at the Arnhem Zoo, in the Netherlands, where I began my career with the other *Pan* species. The German professor would love chimpanzees, since males rule supremely and are constantly vying for position, so much so that I wrote an entire book, *Chimpanzee Politics*, on their schmoozing and scheming. As a student, I began to read Niccolò Machiavelli to gain insights that biology textbooks couldn't offer me. One of the central male characters of that tumultuous period, now four decades in the past, was murdered while I was still there, an event that continues to haunt me, not least because of the gruesome removal of his testicles by the attackers. The other male characters have all died over the years, but the colony still includes their adult sons, who not only look unnervingly like their fathers but also sound like them when they hoot or scream. Chimps have distinctive voices: I used to be able to tell all twenty-five of them apart by their calls alone. I feel very much at home with these primates and consider them absolutely fascinating, but I never have any illusions about how "nice" they are, even if they look like it to most people. They take their power games very seriously and are ready to kill their rivals. That they sometimes kill humans, or bite off their face, as has happened with pet chimps in the United States, is what you can expect if you keep a wild animal in a situation in which sexual jealousy and its dominance drive risk being aroused by our own feeble species. A single adult male chimpanzee has such muscle power (not to mention his daggerlike canine teeth and four "hands") that even a team of five hefty men would never be able to hold him down. Chimps raised around people know this all too well.

The females I knew in Arnhem are still around, however, especially the impressive matriarch of the colony, named Mama. She was never like a bonobo matriarch, who rules the place, but has been alpha among the females for as long as I remember. In her heyday, Mama was an active player in male power struggles. She would rally female support for one male or another, who would be in her debt if he managed to get to the top. This male would do well to stay on her good side, because if Mama turned against him, his career might be over. Mama went so far as to punish females who dared side with males she did not approve of, acting like a party whip. Chimpanzee males physically dominate females, but it is not as if females know nothing about politics or stay out of it. Females in wild communities often do, but on the island in Arnhem this isn't an option. The result is a reduced power gap between the sexes. Since all females are present all the time, actively supporting each other, it is impossible for any male to get around the female power block.

I have always been close to Mama, who greets me with a mixture of respect and affection each time she sees me. She did so already all those years ago, and still does so each time she detects my face in a crowd of visitors. I have been to the zoo every couple of years, and sometimes engage in a bit of friendly grooming with her, but this time I arrived with almost one hundred people in tow, attendees of a symposium at the zoo's convention center. As we walked up to the chimp island, both Mama and another old female, Jimmy, hurried forward to greet me: they gave a series of low grunts, and Mama stretched out a hand to me from a distance. Females typically use this "come here" gesture when they are about to move and want their offspring to jump on their back. I made the same gesture back at her and later helped the caretaker feed the chimpanzees by throwing fruit across the water moat, making sure that Mama, who walks slowly and isn't as skilled as the others at plucking flying oranges out of the air, got enough.

Jealousy was on display right then and there, because Mama's adult daughter, Moniek, snuck up on us to lob a heavy stone from a distance of about forty feet. Moniek's parabolic launch would have hit me in the head had I not kept an eye on her. I caught the rock in the air. Moniek was born while I still worked at the zoo, and I have seen many times how she hates her mother's attention for me. She probably doesn't remember me, hence has no clue why Mama greets this stranger like an old friend. Better throw something at him! Since aimed throwing is viewed by some scholars as a human specialization related to language evolution, I have invited proponents of this theory to experience firsthand what chimps are capable of, but never had any volunteers. Perhaps they realize that stones may be replaced by smelly body products.

Moved by the reunion between Mama and myself, the symposium participants wondered how well chimpanzees recognize us and how well we recognize them. For me, ape faces are as distinct as human faces, even though both species have a bias for their own kind. This bias was ignored not too long ago when only humans were considered good at face recognition. Apes had done poorly on the same tests as applied to humans with the same stimuli, which meant that the apes had been tested on *human* faces. I call this the "anthropocentric bias" in ape research, which is responsible for much misinformation. When one of my co-workers in Atlanta, Lisa Parr, used the hundreds of photographs I had shot in Arnhem to test chimpanzees on portraits of their *own* species, they excelled at it. Seeing the portraits on a computer screen, they were even able to tell which juveniles were offspring of which females, doing so without personally knowing the pictured chimps. In the same way, leafing through a photo album, we can tell from the faces alone which humans are blood relatives.

We live in a time of increasing acceptance of our kinship with the apes. True, humanity never runs out of claims of what sets it apart, but it is a rare uniqueness claim that holds up for over a decade. If we consider our species without letting ourselves be blinded by the technical advances of the last few millennia, we see a creature of flesh and blood with a brain that, albeit three times larger than a chimpanzee's, doesn't contain any new parts. Even our vaunted prefrontal cortex turns out to be of rather typical size compared with that of other primates. No one doubts the superiority of our intellect, but we have no basic wants or needs that are not also present in our close relatives. Just like us, monkeys and apes strive for power, enjoy sex, want security and affection, kill over territory, and value trust and cooperation. Yes, we have computers and airplanes, but our psychological makeup remains that of a social primate.

This is why we had an entire symposium at the zoo on what health care professionals and social scientists might learn from primatology. I was the primatologist, of course, but learned something myself from a discussion on the side. We were talking about where morality gets it justification. If the weight behind it doesn't come from above, who or what provides it? A colleague noted that while the Dutch had become quite secular over the past few decades, there is a growing problem with moral authority. No one publicly corrects anyone anymore, and people have become less civilized as a result. I saw heads nodding around the table. Was this just a frustrated rant by the older generation, always ready to complain about the younger one? Or was there a pattern? Secularization is all around us in Europe, but its moral implications are poorly understood. Even the German political philosopher Jürgen Habermas—an atheist Marxist if there ever was one—has come to regard the loss of religion as perhaps not altogether beneficial, stating that "something was lost when sin became guilt."[8]

The Atheist Dilemma

I am not convinced that morality needs to get its weight from above, though. Can't it come from within? This would certainly work for compassion, but perhaps also for our sense of fairness. A few years ago, we demonstrated that primates will happily perform a task for cucumber slices until they see others getting grapes, which taste so much better. The cucumber eaters become agitated, throw down their veggies, and go on strike. A perfectly fine food has become unpalatable as a result of seeing a companion get something better. We labeled it *inequity aversion*, a topic since investigated in other animals, including dogs. A dog will repeatedly perform a trick without rewards, but refuse as soon as another dog gets pieces of sausage for the same trick.

Such findings have implications for human morality. According to most philosophers, we reason ourselves toward moral truths. Even if they don't invoke God, they're still proposing a top-down process in which we formulate the principles and then impose them on human conduct. But do moral deliberations really take place at such an elevated plane? Don't they need to be anchored in who and what we are? Would it be realistic, for example, to urge people to be considerate of others if we didn't already have a natural inclination to be so? Would it make sense to appeal to fairness and justice if we didn't have powerful reactions to their absence? Imagine the cognitive burden if every decision we took had to be vetted against handed-down logic. I am a firm believer in David Hume's position that reason is the slave of the passions. We started out with moral sentiments and intuitions, which is also where we find the greatest continuity with other primates. Rather than having developed morality from scratch through rational reflection, we received a huge push in the rear from our background as social animals.

At the same time, however, I am reluctant to call a chimpanzee a "moral being." This is because sentiments do not suffice. We strive for a logically coherent system and have debates about how the death penalty fits arguments for the sanctity of life, or whether an unchosen sexual orientation can be morally wrong. These debates are uniquely human. There is little evidence that other animals judge the appropriateness of actions that do not directly affect themselves. The great pioneer of morality research, the Finnish anthropologist Edward Westermarck, explained that moral emotions are disconnected from one's immediate situation. They deal with good and bad at a more abstract, disinterested level. This is what sets human morality apart: a move toward universal standards combined with an elaborate system of justification, monitoring, and punishment.

At this point, religion comes in. Think of the narrative support for compassion, such as the parable of the good Samaritan, or the challenge to our sense of fairness, such as the parable of the workers in the vineyard with its famous conclusion "The last will be first, and the first will be last." Add to this an almost Skinnerian fondness of reward and punishment—from the virgins to be met in heaven to the hellfire awaiting sinners—and the exploitation of our desire to be "praiseworthy," as Adam Smith called it. Humans are in fact so sensitive to public opinion that we only need to see a picture of two eyes glued to the wall to respond with good behavior. Religion understood this long ago and uses the image of an all-seeing eye to symbolize an omniscient God.

But even assigning such a modest role to religion is anathema for some. Over the past few years, we have gotten used to a strident atheism arguing that God is not great (Christopher Hitchens) or is a delusion (Richard Dawkins). The neo-atheists call themselves "brights," thus implying that believers are not as bright. They have replaced Saint Paul's view that nonbelievers live in darkness by its opposite: nonbelievers are the only ones to have seen the light. Urging trust in science, they wish to root ethics in the naturalistic worldview. I do share their skepticism regarding religious institutions and their "primates"—popes, bishops, megapreachers, ayatollahs, and rabbis—but what good could possibly come from insulting the many people who find value in religion? And more pertinently, what alternative does science have to offer? Science is not in the business of spelling out the meaning of life and even less in telling us how to live our lives. The British philosopher John Gray put it as follows: ". . . science is not sorcery. The growth of knowledge enlarges what humans can do. It cannot reprieve them from being what they are."[9] We scientists are good at finding out why things are the way they are, or how they work, and I do believe that biology helps us understand why morality looks the way it does. But to go from there to offering moral advice is a stretch.

Even the staunchest atheist growing up in Western society cannot avoid having absorbed the basic tenets of Christianity. The increasingly secular northern Europeans, whose cultures I know firsthand, consider themselves largely Christian in outlook. Everything humans have accomplished anywhere—from architecture to music, from art to science—developed hand in hand with religion, never separately. It is impossible, therefore, to know what morality would look like without religion. It would require a visit to a human culture that is not now and never was religious. That such cultures do not exist should give us pause.

Bosch struggled with the same issue—not with being an atheist, which was not an option, but with science's place in society. The little figures in his paintings with inverted funnels on their heads or the background buildings in the form of distillation bottles and furnaces reference chemical equipment. However we view science now, it is good to realize that it didn't start out as a very rational enterprise. Alchemy was gaining ground in Bosch's days, yet mixed with the occult and full of charlatans and quacks, which the painter depicted with great humor in front of their gullible audiences. Alchemy turned into

Bosch's paintings abound with references to alchemy, the mystic forerunner of chemistry. *The Garden*'s most recognizable figure—known as the "egg man" or "tree man"—carries a carrousel on his head with a smoking bagpipe-like contraption commonly used as an alchemical vessel.

empirical science only when it liberated itself from these influences and developed self-correcting procedures. But how science might contribute to a moral society remained unclear.

Other primates, of course, have none of these problems, but even they strive for a certain kind of society. In their behavior, we recognize the same values we pursue ourselves. For example, female chimpanzees have been seen to drag reluctant males toward each other to make up after a fight, while removing weapons from their hands. Moreover, high-ranking males regularly act as impartial arbiters to settle disputes in the community. I take these hints of *community concern* as a sign that the building blocks of morality are older than humanity, and that we don't need God to explain how we got to where we are today. On the other hand, what would happen if we were to excise religion from society? I have trouble seeing how science and the naturalistic worldview could fill the void and become an inspiration for the good.

At the end of my weeklong transatlantic excursion, I found time on the plane back to read through the nearly seven hundred responses generated by my blog *Morals without God?* Most comments were constructive and supportive, expressing belief in shades of gray when it comes to the origins of morality. But atheists couldn't resist the occasion to make more digs at religion, thus bypassing my intentions. For me, understanding the need for religion is a far superior goal to bashing it. The central issue of atheism, which is the (non)existence of God, strikes me as monumentally uninteresting. What do we gain by getting in a tizzy about the existence of something no one can prove or disprove? In 2012, Alain de Botton raised hackles by opening his book *Religion for Atheists* with the line "The most boring and unproductive question one can ask of any religion is whether or not it is *true*—in terms of being handed down from heaven to the sound of trumpets."[10] Yet, for some this remains the only issue they can talk about. How did we reach this small-mindedness, as if we've joined a debating club, where all one can do is win or lose?

Science isn't the answer to everything. As a student, I learned about the "naturalistic fallacy" and how it would be the zenith of arrogance for scientists to think that their work could illuminate the distinction between right and wrong. This was not long after World War II, mind you, which had brought us massive evil justified by a scientific theory of self-directed evolution. Scientists had been very much involved in the genocidal machine, conducting unimaginable experiments. Children had been sown together to create conjoined twins, live humans had been operated on without anesthesia, and limbs and eyes had been surgically relocated on people's bodies. I have never forgotten this dark postwar period, during which every scientist who spoke with a German accent was suspect. American and British scientists were not innocent, however, because they were the ones who earlier in the century had brought us eugenics. They advocated racist immigration laws and forced sterilization of the deaf, blind, mentally ill, and physically impaired, as well as criminals and members of minority races. Surgeries to this effect were secretly performed on victims visiting the hospital for other reasons. For those who do not wish to blame this sordid history on science, and prefer to speak of pseudoscience, it will be good to consider that eugenics was a serious academic discipline at many universities. By 1930, institutes devoted to it existed in England, Sweden, Switzerland, Russia, America, Germany, and Norway. Its theories were supported by prominent figures, including American presidents. Its founding father, the British anthropologist and polymath Sir Francis Galton, became a fellow of the Royal Society and was knighted well after having espoused ideas about improving the human race. Notably, Galton felt that the average citizen was "too base for the everyday work of modern civilization."[11]

It took Adolf Hitler and his henchmen to expose the moral bankruptcy of these ideas. The inevitable result was a precipitous drop of faith in science, especially biology. In the 1970s, biologists were still commonly equated with fascists, such as during the heated protest against "sociobiology." As a biologist myself, I am glad those acrimonious days are over, but at the same time I wonder how anyone could forget this past and hail science as our moral savior. How did we move from deep distrust to naïve optimism? While I do welcome a science of morality—my own work is part of it—I can't fathom calls for science to *determine* human values (as per the subtitle of Sam Harris's *The Moral Landscape*).[12] Is pseudoscience something of the past? Are modern scientists free from moral biases? Think of the Tuskegee syphilis study of just a few decades ago, or the ongoing involvement of medical doctors in prisoner torture at Guantánamo Bay.[13] I am profoundly skeptical of the moral purity of science, and feel that its role should never exceed that of morality's handmaiden.

The confusion seems to stem from the illusion that all we need for a good society is more knowledge. Once we have figured out the central algorithm of morality, so the thinking goes, we can safely hand things over to science. Science will guarantee the best choices. This is a bit like thinking that a celebrated art critic must be a great painter or a food critic a great chef. After all, critics offer profound insights in regard to particular products. They possess the right knowledge, so why not let them handle the job? A critic's specialty, however, is post hoc evaluation, not creation. And creation takes intuition, skill, and vision. Even if science helps us appreciate how morality works, this doesn't mean it can guide it anymore than that someone who knows how eggs should taste can be expected to lay one.

The view of morality as a set of immutable principles, or laws, that are ours to discover ultimately comes from religion. It doesn't really matter whether it is God, human reason, or science that formulates these laws. All of these approaches share a top-down orientation, their chief premise being that humans don't know how to behave and that someone must tell them. But what if morality is created in day-to-day social interaction, not at some abstract mental level? What if it is grounded in the emotions, which most of the time escape the neat categorizations that science is fond of? Since the whole point of my book is to argue a bottom-up approach, I will obviously return to this issue. My views are in line with the way we know the human mind works, with visceral reactions arriving before rationalizations, and also with the way evolution produces behavior. A good place to start is with an acknowledgment of our background as social animals, and how this background predisposes us to treat each other. This approach deserves attention at a time in which even avowed atheists are unable to wean themselves from a semireligious morality, thinking that the world would be a better place if only a white-coated priesthood could take over from the frocked one.

Chapter 2

GOODNESS EXPLAINED

The social instincts lead an animal to take pleasure in the society of its fellows, to feel a certain amount of sympathy with them, and to perform various services for them.

—Charles Darwin[1]

Amos was one of the handsomest males I have known, except perhaps on the day he stuffed two entire apples in his mouth, which taught me again that chimpanzees can do things we can't. He had large eyes in a friendly, symmetrical face, a full, shiny coat of black hair, and well-defined muscles on arms and legs. He was never overly aggressive, as some males can be, yet supremely self-confident during his heyday. Amos was beloved. Some of us cried when he died, and his fellow apes were eerily silent for days. Their appetite took a hit.

At the time, we didn't know what his problem was, but we learned postmortem that in addition to a hugely enlarged liver that took up most of his abdomen, he had several cancerous growths. He had lost 15 percent of his weight from a year before, but even though his condition must have been building for years, he had acted normally until his body just couldn't hold out any longer. Amos must have felt miserable for months, but any sign of vulnerability would have meant loss of status. Chimps seem to realize this. A limping male in the wild was seen to isolate himself for weeks to nurse his injuries, yet would show up now and then in the midst of his community to give a charging display full of vigor and strength, after which he'd withdraw again. That way, no one would get any ideas.

Amos didn't betray his condition until the day before his death, when we found him panting at a rate of sixty breaths per minute, with sweat pouring from his face, sitting on a burlap sack in one of the night cages while the other chimps were outside in the sun. Amos refused to go out, so we kept him separate until a veterinarian could take a closer look. The other chimps kept returning indoors to check on him, though, so we cracked open the door behind which Amos sat to permit contact. Amos placed himself right next to the opening, and a female, Daisy, gently took his head to groom the soft spot behind his ears. Then she started pushing large amounts of excelsior through the crack. This is a wood shaving that chimps love to build nests with. They arrange it all around them and sleep on it. After Daisy had given Amos the wood wool, we saw a male do the same. Since Amos was sitting with his back against the wall and not doing much with the excelsior, Daisy reached in several times to stuff it between his back and the wall.

This was remarkable. Didn't it suggest that Daisy realized that Amos must be uncomfortable and that he would be better off leaning against something soft, similar to the way we arrange pillows behind a patient in the hospital? Daisy probably extrapolated from how *she* feels leaning against excelsior, and indeed she is known among us as an "excelsior maniac" (instead of sharing the stuff, she normally hogs it). I am convinced that apes take the perspective of others, especially when it comes to friends in trouble. True, when such abilities have been tested in the laboratory, they have not always been confirmed, but those studies typically ask apes to understand *humans* in some artificial setting. I have already mentioned the anthropocentric bias of our science. In ape-to-ape tests of perspective taking, chimpanzees have fared considerably better, and in the wild they attend to what others know or don't know.[2] We shouldn't be surprised, therefore, that Daisy seemed to grasp Amos's situation.

The next day, Amos was put to sleep. There was no hope for survival, only the certainty of more pain. The incident illustrated two contrasting sides of primate social life. First, primates live in a cutthroat world, which forces a male to conceal physical impairment for as long as possible in order to keep up a tough façade. But second, they are part of a tight community, in which they can count on affection and assistance from others, including nonrelatives. Such dualities are hard to grasp. Popular writers prefer to simplify things by describing the lives of chimpanzees either in Hobbesian terms, as nasty and brutish, or by stressing their friendly side, but in fact it's never one or the other. It's always both. If people ask how chimpanzees can possibly be called empathic, knowing that they sometimes kill one another, my return question is always whether by the same token we shouldn't abandon the whole notion of human empathy as well.

This duality is crucial. Morality would be superfluous if we were universally nice. What would there be to worry about if all that humans ever did was show sympathy for one another, and never steal, never stab someone in the back, never covet another's wife? This is clearly not how we are, and it explains the need for moral rules. On the other hand, we could design a zillion rules to promote respect and care for others, but they'd come to naught if we didn't already lean in that direction. They would be like seeds dropped onto a glass plate: without a chance of taking root. What permits us to tell right from wrong is our ability to be both good and bad.

Daisy's assistance to Amos officially ranks as "altruism," defined as behavior that costs you something (such as taking a risk or expending energy) while it benefits another. Most biological discussions of altruism don't concern themselves much with motives, however, only with how such behavior affects others and why evolution might have produced it. Even though this debate is over 150 years old, it took center stage in the past few decades.

The Gene's Point of View

"Secure your own oxygen mask before assisting anyone else," we are urged at the beginning of every flight. Altruism requires that we take care of ourselves first, which is exactly what one of the chief theoreticians of this field tragically failed to do, as is described by the Israeli science historian Oren Harman in his enthralling book *The Price of Altruism.*

George Price was an eccentric American chemist, who moved to London in 1967, where he became a population geneticist trying to solve the mystery of altruism with brilliant mathematical formulas. He had trouble solving his own problems, though. He had shown little sensitivity to others in his previous life (he abandoned his wife and daughters and was a lousy son to his aging mother), and the pendulum now swung to the other extreme. From a staunch skeptic and atheist, he turned into a devout Christian who dedicated his life to the city's vagabonds. He gave up all of his possessions while neglecting himself. By the age of fifty, he was sinewy and gaunt like an old man, with rotting teeth and a raspy voice. In 1975, Price ended his life with a pair of scissors.

Following a long-standing tradition, Price loved to pit altruism against selfishness. The sharper the contrast, the deeper the riddle of where altruism might have come from. There is of course no shortage of such puzzles. Defending their hive, honeybees die shortly after having stung intruders. Chimpanzees rescue each other from leopard

Elephant altruism on the Kenyan plains. With her tusks, Grace (right) lifted the fallen three-ton Eleanor to her feet, then tried to get her to walk by pushing her. But Eleanor fell again and eventually died, leaving Grace vocalizing with streaming temporal glands—a sign of deep distress. Being matriarchs of different herds, these two elephants were likely unrelated.

attacks. Squirrels give alarm calls that warn others of danger. Elephants try to lift up fallen comrades. But why would any animal act on behalf of another? Isn't this against the laws of nature?

Scientists passionately worked on, debated, and bickered about a theoretical problem that outsiders find esoteric but that lies at the core of recent progress in behavioral biology and evolutionary psychology. Apart from the drama of Price's life and death, there was no lack of momentous events and personal encounters, such as the sublime irony that John Maynard Smith, a famous British evolutionary biologist, brought the even more famous J. B. S. Haldane on his deathbed a book that claimed that birds prevent overpopulation by curtailing their own reproduction. For a biologist, this would count as altruism since it would permit others to reproduce at a cost to oneself. This whole idea received heaps of ridicule in years to follow, however, given how unlikely it is that animals put the greater good before their own. Haldane immediately saw the problem, telling his visitors with a mischievous smile,

Well, there are these blackcock, you see, and the males are all strutting around, and every so often, a female comes along, and one of them mates with her. And they've got this stick, and every time they mate with a female, they cut a little notch in it. And when they've cut twelve notches, if another female comes along, they say, "Now, ladies, enough is enough!"[3]

Popularizers often stress how traits help the survival of the species or the group, yet most biologists—including myself—recoil from evolutionary scenarios that stress the group level. This is because most groups don't act like genetic units. In the primates, for example, virtually all members of a given sex (males in most monkeys, but females in apes) leave their group at puberty to join the neighbors, just as humans frequently intermarry between tribes. This seriously blurs the kinship lines. Primate groups are too genetically "leaky" for natural selection to get a grip on them. The only units that may qualify are those based on shared genes, such as extended families. Haldane was one of the main architects of the "gene's-eye view" of evolution. Looked at from the standpoint of the genes, altruism gets a special meaning. Even if one loses one's own life to save a relative, one still perpetuates genes that one has in common with this relative. Helping kin is therefore like helping oneself. Stooped drunk over a beer, Haldane is said to have slurred, "I'll jump into a river for two brothers and eight cousins," thus foreshadowing the theory of kin selection proposed by William Hamilton, one of the brightest and nicest biologists since Charles Darwin.

I add "nicest" to contrast Hamilton with the scientist who in fact coined the term "kin selection" in an article that ran off with Hamilton's idea without offering him credit. This was the same Maynard Smith mentioned before, who upon hearing of Hamilton's idea is said to have exclaimed, "Of course, why didn't I think of that!"[4] Ever since Hamilton found out which anonymous reviewer on his seminal paper had delayed its publication, he harbored a burning grudge against Maynard Smith despite the many apologies he received. Price almost suffered the same fate when Maynard Smith merely wanted to thank him for his ideas on restrained combat ("Why don't venomous snakes use their deadly fangs on each other?"), but fortunately Price managed to gain co-authorship instead.

Initially, kin selection overshadowed every discussion of altruism, owing to a focus on social insects, such as bees and termites, which live in colonies of close relatives. But a second explanation gained equal prominence. Robert Trivers, an American evolutionary biologist, proposed that cooperation among nonrelatives often relies on reciprocal altruism: helpful acts that are costly in the short run nevertheless produce long-term benefits if favors are being returned. If I rescue a friend who almost drowns, and he rescues me under similar circumstances, both of us will be better off than we would have been on our own. Reciprocal altruism allows cooperative networks to expand beyond kinship ties.

Not surprisingly, most players in this long quest were politically opinionated. One of them, the British statistician and biologist Ronald Fisher, was an avowed eugenicist, who felt that the human race could use some genetic improvement. Another one, the Hungarian American game theoretician John von Neumann, was so enthralled by his own calculations that he urged the U.S. Senate in 1955 to drop the atomic bomb onto the Soviets, saying, "If you say why not bomb them tomorrow, I say why not today?"[5] Others were card-carrying Communists, however, and early on there was, of course, Petr Kropotkin, the Russian anarchist prince. Despite the charge by opponents that evolutionary biology is a right-wing plot, the altruism debate took place mostly on the left rather than right side of the ideological spectrum. I know this firsthand, since Trivers and the late Hamilton occasionally met in California at meetings of the Gruter Institute for Law and Behavioral Research, of which I have been part for decades. I once conducted an interview with Bob to ask him about the implications of his theory. Here is one of his answers:

> FdW: *Reading between the lines, I recognize in your paper the same sort of social commitment that led Kropotkin to develop his ideas. . . .*
>
> Trivers: *You're right about my political preferences. When I left mathematics, and cast about what I was going to do in college, I said (self-ironic bombast): "All right, I'll become a lawyer and fight for civil rights and against poverty!" Someone suggested that I take up U.S. history, but you know at that time, in the early 1960s, their books were entirely self-congratulatory. I ended up in biology.*
>
> *Because I remained a political liberal, for me, emotionally, to see that just pursuing this scratch-my-back argument would generate rather quickly a reason for justice and fairness was very gratifying because it was on the other side of the fence of that awful tradition in biology of the right of the strongest.*[6]

After having attended Price's funeral, Hamilton went to rescue the dead scientist's papers in a squatter's flat, commenting how much kinship he felt with the man who by the end of his life was talking to God ("I try to be in everything a slave of the Lord, and bring large and small matters to Him for decision").[7] One explanation of Price's untimely death is that he was deeply upset by the unpleasant implications of his own calculations, such as the impossibility to evolve loyalty to the in-group without also evolving torture, rape, and murder of the out-group. He despaired at the thought that altruism might not have come into existence without its negative flip side. But Price also harbored the idea that self-interest stands in the way of genuine altruism. This giant misunderstanding may have cost him his life as he probed the boundaries of human nature by testing his own capacity for self-sacrifice. Never mind that most human altruism doesn't operate this way. It grows out of empathy with those in need, and the whole point of empathy is a blurring of the line between self and other. This obviously makes the difference between selfish and unselfish motives rather hazy.

Realizing that empathy needs to be part of any theory of human altruism, Trivers tested the concept on Hamilton: "Long ago I spoke to Bill about it, I said 'What about empathy Bill?' and he said, 'What's empathy?' As if it didn't exist, as if there was no such thing."[8] Too bad, I should add, because attention to the way altruism *works* might have reduced the amount of controversy and confusion about what genes are capable of. The road between genes and behavior is far from straight, and the psychology that produces altruism deserves as much attention as the genes themselves. The error of early theorizing was to skip these complexities.

Empathy is mostly a mammalian trait, so the deeper error was that great thinkers had lumped all sorts of altruism together. There are the bees dying for their hive and the millions of slime mold cells that build a single, sluglike organism that permits a few among them to reproduce. This kind of sacrifice was put on the same level as the man jumping into an icy river to rescue a stranger or the chimpanzee sharing food with a whining orphan. From an evolutionary perspective, both kinds of helping are comparable, but psychologically speaking they are radically different. Do slime molds even have motivations the way we do? And when bees sting an intruder aren't they driven by aggression rather than by the benign motives we associate with altruism? Mammals have what I call an "altruistic impulse" in that they respond to signs of distress in others and feel an urge to improve their situation. To recognize the need of others, and react appropriately, is really not the same as a preprogrammed tendency to sacrifice oneself for the genetic good.

With the increasing popularity of the gene's-eye view, however, these distinctions were overlooked. This led to a cynical outlook on human and animal nature. The altruistic impulse was downplayed, ridiculed even, and morality was taken off the table entirely. We were only slightly better than social insects. Human kindness was seen as a charade and morality as a thin veneer over a cauldron of nasty tendencies. This outlook, which I have dubbed *Veneer Theory*, can be traced back to Thomas Henry Huxley, also known as "Darwin's Bulldog."

The Bulldog's Cul-de-Sac

Darwin being defended by Huxley is a bit like Albert Einstein being defended by me. For the life of me, I can't grasp the theory of relativity. I am just not mathematically inclined enough, even though I can follow the speeding-train example and other simplifications for dummies. All that Einstein could possibly expect from me is an arm-waving story about how I think he had a great idea that superseded Isaac Newton's mechanistic notions. Not much help, of course, but not unlike what happened to Darwin when Huxley took up his cause. Huxley lacked formal education and was a self-taught comparative anatomist of great standing. He was notoriously reluctant, however, to accept natural selection as the chief engine of evolution and also had trouble with gradualism. These are no minor details, which is why we shouldn't be surprised that one of last century's leading biologists, Ernst Mayr, harshly concluded that Huxley "did not represent genuine Darwinian thought in any way."[9]

Huxley is best known for wiping the floor with Bishop "Soapy Sam" Wilberforce in an 1860 public debate about evolution in which the bishop mockingly inquired whether Huxley descended from the primates through his grandfather's or his grandmother's side. Huxley is said to have replied that he wouldn't mind to be the progeny of an ape, but would be quite ashamed to be connected to someone who abuses his oratorical gifts to obscure the truth.

It is good to take this famous story with a grain of salt, though.

Apart from the fact that it was concocted decades after the encounter, there is the problem that Huxley's voice was too weak to command an audience. In the days before microphones, this mattered. One of the other scientists present wrote dismissively that Huxley was unable to throw his voice over so large an assembly and that it was he, the botanist Joseph Hooker, who had confronted the bishop: "I smashed him amid rounds of applause. I hit him in the wind at the first shot in ten words taken from his own ugly mouth." The bishop himself, by the way, felt that he had thoroughly taken care of his opponents.[10]

This must have been one of those rare debates where everyone was a winner! It was Huxley, however, who went down in history as the great defender of science against religion. Being a far more retiring person, Darwin needed a soldier by his side to make the case for his controversial ideas. Huxley loved to fight and had been looking for a cause to sink his teeth into. After having read *The Origin of Species*, he

Thomas Henry Huxley, the combative public defender of Darwin, called himself "agnostic." He had a strong religious bent, however, that kept him from agreeing with Darwin on the issue of moral evolution, which he considered an impossibility (cartoon by Carlo Pellegrini in *Vanity Fair* of 28 January 1871).

eagerly offered Darwin his services: "I am sharpening up my claws & beak in readiness."[11]

Contributing to Huxley's reputation as the slayer of religion is his invention of the term "agnostic," meaning that he wasn't sure of God's existence. He saw agnosticism as a method, however, not a creed. He advocated scientific arguments that were based purely on evidence rather than on any higher authority, a position nowadays known as "rationalist." Huxley deserves admiration for this major step in the right direction, but the great irony is that he remained deeply religious and allowed this to color his outlook. He described himself as a "scientific Calvinist," and much of his thinking followed the somber, joyless precepts of the doctrine of original sin. Given that pain in the world is a certainty, he said, we can only hope to endure it with clenched teeth—his "grin and bear it philosophy."[12] Nature lacks the capacity of producing any good, or in Huxley's own words:

> *The doctrines of predestination, of original sin, of the innate depravity of man and the evil fate of the greater part of the race, of the primacy of Satan in this world, of the essential vileness of matter, of a malevolent Demiurgus subordinate to a benevolent Almighty, who has only lately revealed himself, faulty as they are, appear to me to be vastly nearer the truth than the "liberal" popular illusions that babies are all born good. . . .*[13]

The primacy of Satan? Does this sound like an agnostic? A book entitled *Lay Sermons* allowed Huxley to compete with sermons from the pulpit. He adopted such a preachy tone that his critics called him self-righteous and full of Puritanical convictions. Huxley's religious attitudes, hidden underneath a desire for objective truth, explain why he developed Veneer Theory and its bleak assessment of human nature. He saw human ethics as a victory over nature similar to a well-tended garden. The gardener struggles every day to keep his garden from going wild. As Huxley put it, the horticultural process is opposed to the cosmic process. Nature tries to undermine the gardener's efforts by invading his plot with despised weeds, slugs, and other pests ready to choke off the exotic plants he wants to cultivate.

This metaphor says it all: ethics is a uniquely human answer to an unruly, wicked evolutionary process. In his well-known lecture on this topic for a large audience in Oxford, in 1893, Huxley summed up his position:

> *The practice of that which is ethically best—what we call goodness or virtue—involves a course of conduct which, in all respects, is opposed to that which leads to success in the cosmic struggle for existence.*[14]

Unfortunately, the anatomist gave no hint where humanity might have unearthed the will and strength to defeat its own nature. If we are indeed devoid of natural benevolence, how and why did we decide to become model citizens? And if doing so was to our advantage, as one would hope, why did nature refuse us a helping hand? Why must we perpetually sweat in the garden to keep our immoral impulses at bay? It is a bizarre theory, if that's what we call it, according to which morality is only an evolutionary afterthought barely capable of concealing the sinners we truly are. Note that this dark idea was entirely Huxley's. I agree with Mayr that it doesn't bear the slightest resemblance to Darwin's thinking. In the words of Huxley's biographer, he "was forcing his ethical Ark against the Darwinian current which had brought him so far."[15]

In the end, Darwin desperately needed a defender against his public defender. He got one in the form of Kropotkin, a first-rate naturalist. Whereas Huxley was a city boy with little firsthand knowledge of noncadaverous animals, Kropotkin had traveled around Siberia and noticed how rarely animal encounters fit the gladiatorial style hyped by Huxley, who imagined a "continuous free fight." Kropotkin had noticed frequent cooperation between members of the same species. Huddling together in the cold or collectively standing up to predators—such as wild horses against wolves—was critical for survival. Kropotkin emphasized these themes in his 1902 book *Mutual Aid*, which was explicitly directed against "infidels" such as Huxley, who misinterpreted Darwin. True, Kropotkin went overboard in the other direction, cherry-picking examples of animal solidarity to support his political views, yet he was right to protest Huxley's depiction of nature, which was poorly informed by reality.

For me, the biggest question of all is how to exit from the Huxleyan cul-de-sac. If we are not allowed to talk about God and if evolution offers no answer either, what could possibly explain human morality? With both religion and biology out of the picture, all I see is a big black hole. And the most astonishing of all is that biologists, like bad drivers, steered us into the same blind alley again a century later.

My Life as a Toilet Frog

In Australia, it is not unusual to find a sizable frog in your toilet bowl. You may try to move it out, but the frog will happily hop back into the bowl, where it attaches itself with its suction cup toes during the occasional tsunamis that we humans produce. These frogs don't seem to mind the body waste swirling down the bowl.

But I do! I felt like a toilet frog during the last three decades of the preceding century. I had to hang on desperately each time a book came out on the human condition, whether written by biologists, anthropologists, or science journalists, because most of them advocated ideas totally anathema to the way I view our species. One can consider humans as either inherently good but capable of evil or as inherently evil yet capable of good. I happen to belong to the first camp, but the literature stressed only the negative side. Even positive traits had to be phrased as if they were problematic. Animals and humans love their families? Let's call it "nepotism." Chimpanzees permit friends to eat food out of their hands? Let's call it "pilfering" and "scrounging." The prevailing tone was full of misgivings about kindness. Here is a characteristic statement, cited over and over in this literature:

> *No hint of genuine charity ameliorates our vision of society, once sentimentalism has been laid aside. What passes for co-operation turns out to be a mixture of opportunism and exploitation. . . . Given a full chance to act in his own interest, nothing but expediency will restrain [a person] from brutalizing, from maiming, from murdering—his brother, his mate, his parent, or his child. Scratch an "altruist" and watch a "hypocrite" bleed.*[16]

Altruists are just hypocrites for Michael Ghiselin, an American biologist so well known for his work on the sea slug that one of its defensive chemicals (*ghiselinin*) was named after him. But instead of dealing with slugs, the above statement was about humans. It set the tone for much that followed, as was also reflected two decades later in *The Moral Animal*, by the science journalist Robert Wright: ". . . the pretense of selflessness is about as much part of human nature as is its frequent absence."[17] And then there is George Williams, the American evolutionary biologist who took perhaps the most extreme position. Offering a dark assessment of nature's "wretchedness," he felt that calling nature "amoral" or "morally indifferent," as Huxley wisely had done, didn't go far enough. He rather accused nature of "gross immorality," thus becoming the first and hopefully last biologist to infuse the evolutionary process with moral agency.[18]

The argument typically ran as follows: (1) natural selection is a selfish, nasty process, (2) this automatically produces selfish and nasty individuals, and (3) only romantics with flowers in their hair would think otherwise. Obviously, Darwin was claimed to be on board with the expulsion of morality from the natural domain, as if Darwin would ever have let himself get stuck in Huxley's cul-de-sac. Darwin was too smart for this, as I will explain below, which is why the height of absurdity was reached when Richard Dawkins explicitly disavowed Darwin, telling an interviewer in 1997 that "in our political and social life we are entitled to throw out Darwinism."[19]

I decline to cite more smelly stuff. The only scientist to reach a perfectly logical conclusion—even if I fully disagree—was Francis Collins, head of America's largest federal research agency, the National Institutes of Health. Having read all those books that doubt the evolution of morality, and observing that humanity nevertheless possesses a measure of it, Collins saw no way around a supernatural source: "The Moral Law stands out for me as the strongest signpost of God."[20]

Naturally, the esteemed geneticist became the laughingstock of the nascent atheist movement. Some claimed he was polluting science with faith, while Dawkins, with characteristic lenience, called him "not a bright guy."[21] Never mind the deeper problem, which is that by seriously fumbling the morality question, biologists had left the door wide open for alternative accounts. The whole episode could have been avoided had Collins encountered a more thoughtful evolutionary literature, one taking its lead from Darwin's *The Descent of Man*. Reading this book, one realizes that there is absolutely no need to throw the old man under the bus. Darwin had no trouble aligning morality with the evolutionary process, and recognizing the human capacity for good. Most interesting for me, he saw emotional continuity with other animals. For Huxley, animals were mindless automata, but Darwin wrote an entire book about their emotions, including their capacity for sympathy. One memorable example was how a particular dog would never walk by a basket where a sick friend lay, a cat, without giving her a few licks. Darwin saw this as a sure sign of affection. In his last note to Huxley, right before his death, Darwin couldn't resist gently poking fun at his friend's Cartesian bent, hinting that if animals are machines, then humans must be, too: "I wish to God there were more automata in the world like you."[22]

Darwin's writing massively contradicts Veneer Theory. He speculated, for example, that morality grew straight out of animal social instincts, saying that "it would be absurd to speak of these instincts as having been developed from selfishness."[23] Darwin saw the potential for genuine altruism, at least at the psychological level. Like most biologists, he drew a sharp line between the *process* of natural selection, which indeed has nothing nice about it, and its many *products*, which cover a wide range of tendencies. He disagreed that a nasty process ipso facto needs to produce nasty results. To think so is what I have dubbed the "Beethoven error," since it is like evaluating Ludwig van Beethoven's music on the basis of how and where it was composed. The maestro's Viennese apartment was a messy, smelly pigsty, strewn with waste and unemptied chamber pots. Of course, no one judges Beethoven's music by it. In the same way, even if genetic evolution proceeds through death and destruction, this doesn't taint the marvels it has produced.

That seems an obvious point to make, but once I had done so at length in my 1996 book *Good Natured*, I grew tired of the battle against Veneer Theory. For three long decades, it was greeted with irrational exuberance, no doubt owing to its simplicity: everyone understood it, everyone loved it. How could I disagree with something so obvious?

But then a curious thing happened: the theory vaporized. Rather than dying from a slow feverish illness, Veneer Theory suffered a massive heart attack. I don't quite understand how and why this happened. Perhaps it was Y2K, but by the end of the twentieth century the need to combat Darwin's "infidels" had quickly evaporated. New data were coming in, first as a trickle, then as a steady stream. Data have this wonderful quality of burying theories. I remember picking up an article in 2001 entitled "The Emotional Dog and Its Rational Tail." It was by Jonathan Haidt, an American psychologist, who argued that we arrive at moral decisions through intuitive processes. We hardly think about them. Haidt presented subjects with stories of odd behavior (such as a one-night stand between a brother and sister), which they immediately disapproved of. He then challenged every single reason his subjects could come up with until they ran out of arguments. They might say that incest leads to abnormal offspring, but in Haidt's story the siblings used effective contraception, which took care of this problem. Most of his subjects quickly reached the stage of "moral dumbfounding": they stubbornly insisted the behavior was wrong without being able to explain why.

Haidt concluded that moral decisions come from the "gut." Our emotions decide, after which human reason does its best to catch up. With this dent in the primacy of logic, Hume's moral "sentiments" made a comeback. Anthropologists demonstrated a sense of fairness in people across the world, economists found humans to be more cooperative and altruistic than the *Homo economicus* view would allow, experiments with children and primates found altruism in the absence of incentives, six-month-old babies were said to know the difference between "naughty" and "nice," and neuroscientists found our brains to be hardwired to feel the pain of others. We had come full circle by 2011, when humans were officially declared "supercooperators."

Every new development slammed another nail in the coffin of Veneer Theory until the common view had swiveled around 180 degrees. It is now widely assumed that we are designed in body and mind to live together and take care of each other, and that humans have a natural tendency to judge others in moral terms. Instead of being a thin veneer, morality comes from within. It's part of our biology, a view supported by the many parallels found in other animals. In a few decades, we had gone from calls to teach our children to be nice, since our species lacks any and all natural inclinations in this direction, to the consensus that we are born to be good and that nice guys finish first.

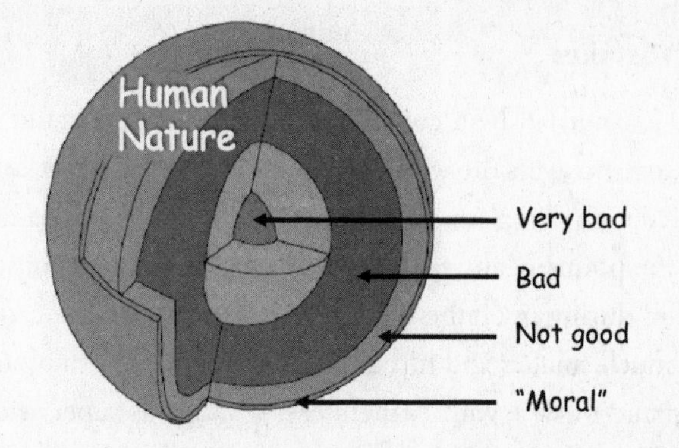

Veneer Theory used to be the dominant biological view of human nature. It regarded genuine kindness as either absent or an evolutionary misstep. Morality was a thin veneer barely able to conceal our true nature, which was entirely selfish. In the past decade, however, Veneer Theory has succumbed to overwhelming evidence for innate empathy, altruism, and cooperation in humans and other animals.

How radically attitudes have changed is clear every time I show audiences the notorious Ghiselin statement that had landed me in the toilet to begin with: "Scratch an altruist and watch a hypocrite bleed." Although I have featured this cynical line for decades in my lectures, it is only since about 2005 that audiences greet it with audible gasps and guffaws as something so outrageous, so out of touch with how they see themselves, that they can't believe it was ever taken seriously. Had the author never had a friend? A loving wife? Or a dog, for that matter? What a sad life he must have lived! Seeing the shocked reaction, and how widespread it has become, I'm left wondering whether my audiences have changed under the influence of new evidence, or whether it's perhaps the other way around. Have we entered a new Zeitgeist, and is science simply catching up?

Regardless, it has left my toilet smelling like a rose. Finally, I can release my grip, stretch my legs, and swim around.

Full of Mistakes

But still, it could all be a colossal mistake. Kindness could be maladaptive, occurring at the wrong time and wrong place. Take the way Daisy cared for a dying Amos, or how humans nurse the terminally ill. What is the point? Many people take care of an aging spouse, as my mother did during my father's last years. What a load she carried, she being so much smaller and he barely able to walk. Or, think of caring for a husband or wife with Alzheimer's, who needs supervision every minute of the day, doesn't appreciate your efforts, and reacts surprised every time you enter the room, complaining that you've abandoned him or her. Stress and exhaustion are all you will ever get out of it. In none of these cases is there much chance of repayment, whereas evolutionary theory insists that altruism should benefit either blood relatives or those willing and able to return the favor. A dying spouse doesn't fit the bill.

Since Daisy, my mother, and millions of caretakers deviate from evolutionary dogma, there has been much talk of "misfiring genes" that cause us to be better than is good for us. Don't be misled by this kind of rhetoric, though. The reason why misfiring genes don't exist as a concept in genetics is that genes are just little chunks of DNA that know nothing and intend nothing. They have the effect they have without any goal in mind, hence are incapable of mistakes. It would be more appropriate to call rampant altruism a glorious accident, but few experts are in a celebratory mood. Their message is rather sour, as if a great theory about the selfish origins of altruism is regrettably spoiled by the facts. They complain that "almost everything in modern life is a mistake from the genes' point of view," but never conclude that this makes their theories largely irrelevant.[24]

There is the mistake of sending money to faraway places struck by a tsunami or an earthquake. Anonymously donating blood is a mistake. Working in a soup kitchen or shoveling snow for an old lady is a mistake, as is pouring all of our resources into an adopted child. The latter is an incomprehensible multiyear mistake made by thousands of families ignorant of the fact that children who don't share their genes lack any value. Families do the same with pets, providing extraordinary care for animals that lack repayment options. Other common mistakes are warning a stranger of danger, pointing out that someone has left his coat in a restaurant, and picking up a stranded driver. Human life is chock-full of mistakes, large and small. The same applies to the lives of other primates.

Take Phineas, Amos's father. Not that Amos knew that Phineas was his father. In chimpanzee society, there are no permanent bonds between males and females, so that in principle any male might be your father. Phineas had been alpha in earlier years, but when he turned forty, he began to take it easy. He loved to play with juveniles, groom with females, and act as policeman. As soon as he heard a squabble, Phineas would go over to make a big show of force, with all his hair on end, to break it up. He would stand between the contestants until the screaming stopped. This "control role" is also well documented for wild chimpanzees. Remarkably, males in this role don't take sides: they defend the weaker party even if the attacker is their best buddy. I have often puzzled over their impartiality, which deviates from so much else that chimpanzees do. By transcending the performer's social biases, the control role truly aims at what's best for the community.

Jessica Flack and I demonstrated how much the group benefits from such behavior by temporarily removing males that act as arbiters. The result was a society coming apart at its seams. Aggression increased and reconciliation decreased. Order was restored, however, as soon as we returned the males to their group.[25] This still leaves the question, however, why they do it. What is in it for them? The main idea is that high-ranking males gain respect and popularity by coming up for the underdog. But while this may be a perfectly good strategy for younger males, I have trouble applying it to Phineas. Toward the end of his life, this gentle old guy was clearly over the hill and seemed to have few ambitions left. Yet, he zealously monitored discord in the group. His push for harmony was to everyone's advantage except perhaps his own. Are chimpanzees, too, more generous than they should be according to gene-centric theories?

It is quite common for chimps to help unrelated individuals, such as when Washoe, the world's first chimpanzee trained in American Sign Language, heard a female she barely knew scream and hit the water. Racing across two electric fences to reach the victim, Washoe dragged her to safety.[26] Another case concerned Tia, a wild female, at Fongoli, in Senegal, who had lost her infant to poachers. Fortunately, researchers managed to confiscate the baby ape to return it to the group. Jill Pruetz describes how Mike, an unrelated adolescent male, who was too young to be the infant's father, picked it up from where the scientists had left it and carried it straight to Tia. He obviously knew to whom the baby belonged and probably also had noticed how much trouble Tia had in moving around after having been mauled by the poachers' dogs. For two days, Mike carried the baby during group travel while Tia limped behind.[27]

Even the costliest investment of all, the adoption of unrelated young, is not unknown. Not just in females, where one might expect it, but a recent report by Christophe Boesch from Ivory Coast lists at least ten wild male chimps who, over a period of three decades, adopted juveniles who had lost their mothers.[28] In 2012, Disneynature released its popular movie *Chimpanzee*, which captured how Freddy, the community's alpha male, took Oscar under his wing. This documentary was based on real events. When Oscar's mother suddenly died of natural causes, the camera crew happened to be at the right place at the right time. It stayed around even though the prospects for little Oscar looked bleak. Freddy followed the pattern of other adoptive males, who shared food with youngsters, allowed them to sleep in their night nests, protected them against danger, and diligently searched for them when lost. Some cared for their charges for over one year, and one male did so for over five years (chimpanzees don't reach adulthood until they are at least twelve). Barring nursing, these stepfathers were taking on the same burden as mothers do for their offspring, strongly enhancing the orphans' chances at survival. According to DNA samples, adoptive males are not necessarily related to their charges. Oscar was lucky.

Rather than concluding that chimps, too, make mistakes, let's move away from such normative language and its implication that we are born to obey our genes. Why not simply recognize the disconnect between the origin of a trait and its current use. Tree frogs evolved suction cups to cling to leaves, yet may use them to survive in a toilet. Primate hands evolved to grasp tree branches, but I play the piano with them, and baby monkeys use them to cling to their mothers. Many traits evolved for one reason yet have come to serve others as well. I have never heard anyone call fingers gliding over a piano "a mistake," so why apply this sort of language to altruism? One might counter that altruism has a cost, whereas piano playing does not, and that this justifies the "mistake" terminology. But how sure are we that generalized empathy and lifelong commitments don't pay off in the long run? I have never seen proof that such behavior harms us, and rather suspect the opposite. Friedrich Nietzsche, who famously gave us the "God is dead" phrase was interested in the sources of morality. He warned that the emergence of something (whether an organ, a legal institution, or a religious ritual) is never to be confused with its acquired purposes: "Anything in existence, having somehow come about, is continually interpreted anew, requisitioned anew, transformed and redirected to a new purpose."[29]

This is a liberating thought, which teaches us to never hold the history of something against its possible applications. Even if computers started out as calculators, that doesn't prevent us from playing games on them. Even if sex evolved for reproduction, anyone is free (up to a point) to engage in it for fun. There is no law that says traits need each and every time to serve the purpose for which they evolved. The same applies to empathy and altruism, which is why we should simply replace the word "mistake" with "potential." Nothing keeps me from empathizing with a stranded whale and joining efforts to haul it back to the ocean, even if human empathy didn't come into existence with whales in mind. I'm just applying my innate empathic capacity to its fullest potential.

Nietzsche was right: an item's history has limited relevance in the here and now. While I stand in awe before the insights provided by Price, Hamilton, Trivers, and others into the evolutionary background of altruism, I see no reason to turn these insights into a dogma about how humans *ought to* behave.

Hedonic Kindness

Science tells us that we breathe in order to supplement our bodies with oxygen. Lacking this knowledge, however, I'd still do exactly the same, like millions of humans before me and billions of animals. Awareness of O_2 is not what drives breathing. Similarly, when biologists speculate that altruism evolved for its payoffs, it doesn't mean that actors need to know about this. Most animals don't think ahead, as in "If I do this for him, he may return the favor tomorrow." Lacking foresight, they just follow a benevolent impulse. The same applies to humans. Except in business or between unacquainted people, humans rarely tally up the costs and benefits of their behavior, especially among friends and family. In fact, doing so is a bad sign, which family therapists use as an indicator that a marriage is on the rocks.

Both human and animal altruism may be genuine, therefore, in that it lacks ulterior motives. This is true to the point that we have trouble suppressing it. James Rilling, an Emory colleague of mine, concluded from neuroimaging experiments that we have "emotional biases toward cooperation that can only be overcome with effortful cognitive control." Think about it: this means that our first impulse is to trust and assist; only secondarily do we weigh the option of *not* doing so, for which we need reasons. This is the exact opposite of being driven by incentives. Only one category of people lacks this natural impulse, which explains my usual quip that Veneer Theory perfectly captures the psychopathic mindset. Rilling further showed that when normal people aid others, brain areas associated with reward are activated. Doing good feels good.

This "warm glow" effect brings a touching image to mind that I have seen countless times while working with rhesus monkeys. The behavior in question was not exactly altruistic, but very close to the source of all mammalian nurturance. Every spring, our zoo troops produced dozens of newborns. The babies held magnetic appeal for juvenile females, who would try to get their little hands on them by patiently grooming their mothers. It would take a long time of hanging around the mother until the baby would be released to take a few wobbly steps toward the would-be sitter. She'd pick it up, carry it around, turn it upside down to inspect its genitals, lick its face, groom it from all sides, but eventually doze off with the baby firmly clutched in her arms. We took bets on how long it would take. Five minutes, ten minutes? The sleepiness that overcame the babysitters gave the impression that they were in a trance, or perhaps ecstatic, having waited so long for their lucky break. As they held their treasure, release of oxytocin in their bloodstream and brains, known as the hormone of love, weighed down their eyelids. Their sleep would never last long, though, and soon they'd return the baby to its mother.

The joy of baby care prepares young females for the most altruistic act of all. Mammalian maternal care is the costliest, longest-lasting investment in other beings known in nature, starting with nourishment of the fetus and ending many years later. Or, as most parents would say, never. Strangely enough, however, maternal care has been largely absent from the altruism debate. Some scientists don't even want to count it as altruism, since it doesn't fit their emphasis on sacrifice. They want to speak of altruism only if it harms the performer, at least in the short run. No one should be eager to be an altruist, let alone take pleasure in it. I call this the *altruism-hurts hypothesis*, which is deeply erroneous. After all, the definition of altruism is not that it needs to cause pain, only that it carries a cost.

Mind you, biologists have absolutely no trouble explaining why a female mammal would care for her offspring. How else is she going to propagate? We also know how much women want babies. I don't want to be grisly about it, but the desire is strong enough that some women kill for it, and cut open another's belly. Or they steal babies from the nursery. These are sick cases, yet illustrate the overwhelming desire, and the reason why baby care isn't regarded as a sacrifice. With maternal care not being much of a puzzle, science has focused on more perplexing behavior. Science seeks challenges. Yet, I would still argue that, at least for mammals, maternal care is the prototypical form of altruism, the template for all the rest. We ignore it at our peril. It is telling that not a single woman scientist that I know of has gotten carried away by the question of where altruism comes from. For women, maternal care would be hard to leave out, as has been illustrated by two women who did write on human cooperation. Sarah Hrdy, an American anthropologist, proposes an "it takes a village" theory according to which the human team spirit started with collective care for the young, not just by mothers but by all adults around. Similarly, Patricia Churchland, an American philosopher well versed in neuroscience, treats human morality as an outgrowth of caring tendencies. The neural circuitry that regulate the organism's own bodily functions has been co-opted to include the needs of the young, treating them almost like extra limbs. Our children are part of us, so we protect and nurse them unthinkingly, the way we do our bodies. The same brain mechanism provides the basis for other caring relations.

This would explain observable sex differences, which start early in life. At birth, girl babies look longer at faces than do boy babies, who look longer at mechanical toys. Later in life, girls are more prosocial than boys, better readers of emotional expressions, more attuned to voices, more remorseful after having hurt someone, and better at taking someone else's perspective. We also have learned that empathy is enhanced by oxytocin sprayed into the nostrils of both men and women, thus fooling them with the maternal hormone par excellence (oxytocin is associated with childbirth and nursing). In our own studies, we have found that female chimpanzees console distressed parties more often than males do. They approach victims of aggression, tenderly put an arm around them, and hug them until the screaming stops. Females are the more nurturing sex.

If maternal care was almost too obvious for theoreticians to consider, it is also the most self-rewarding care, which brings me to the *altruism-feels-good hypothesis.* Invariably, nature associates things that we need to do with pleasure. Since we need to eat, the smell of food makes us drool like Pavlov's dogs, and food consumption is a favorite activity. We need to reproduce, so sex is both an obsession and a joy. And to make sure we raise our young, nature gave us attachments, none of which exceeds that between mother and offspring. Like any other mammal, we are totally preprogrammed for this in body and mind. As a result, we barely notice the daily efforts on behalf of our progeny and joke about the arm and leg that it costs. Distant relatives and nonrelatives obviously recruit less help, but the underlying satisfaction remains the same, an insight already present in *The Meditations* of the second-century Roman emperor Marcus Aurelius (". . . acts that are consistent with nature, like helping others, are their own reward").[30] We are group animals, who rely on each other, need each other, and therefore take pleasure in helping and sharing.

In the 1996 movie *Marvin's Room*, Bessie (played by Diane Keaton) is visited by her more worldly sister (Meryl Streep). Bessie devoted many years of her life caring for their father, never getting any help from her sister. When Bessie tells her sister that she feels she has been lucky to have had her parents, having so much love in her life, her sister misunderstands, in her own self-centered way, what she means, and tells her, "They love you very much." Bessie corrects her, saying, "That's not what I mean, no. I mean that I have been so lucky to be able to love someone so much." Altruism can fill us with happiness.

The odd notion that altruism has to hurt drove George Price to try his hand at extreme self-sacrifice. One can't be a great altruist without suffering, he thought, and so he gave up all of his possessions and neglected himself to the point that he became miserable. He didn't realize that self-neglect is counterproductive, a theme well known to charity workers. As with the oxygen masks on airplanes, the self needs to be fed before it can take care of anyone else. I have often thought about where the odd altruism-hurts notion may have come from. It's totally anathema, for example, to Buddhism, in which compassion for others is supposed to fill us with joy. This effect isn't limited to self-reflecting adults, but occurs also in toddlers, who derive greater satisfaction from giving treats to others than from receiving them.[31] There is also intriguing evidence from a study of people taking care of a sick spouse or parent. The psychologist Stephanie Brown found that caregivers barely notice the cost of their behavior. They feel at one with their charges and derive such great fulfillment from being needed that they live longer than individuals without a need to take care of others.

On the basis of my personal experience taking care of my wife, whose life was imperiled by breast cancer, I totally agree that the word "sacrifice," although often applied, misses the point. Nothing comes more naturally to us than taking care of loved ones. Churchland correctly saw continuity between caring for one's own body, caring for one's children, and caring for those close to us. Our brain has been designed to blur the line between self and other. It is an ancient neural circuitry that marks every mammal, from mouse to elephant. In a Thai nature reserve, I encountered a blind elephant walking around with her seeing-eye friend. The two unrelated females appeared joined at the hip. The blind one depended on the other, who seemed to understand this. As soon as the latter moved away, one could hear deep rumbling sounds coming from both of them, sometimes even trumpeting, which indicated the other's whereabouts to the blind elephant. This noisy spectacle would continue until they were reunited again. An intensive greeting followed, with lots of ear flapping, touching, and mutual smelling. They enjoyed a close friendship, which enabled the blind female to lead a reasonably normal elephant life.

Given its intrinsic rewards, some like to label care for family and close associates "selfish," at least at an emotional level. While not incorrect, this obviously undermines the whole distinction between selfishness and altruism. If my eating all the food on the table is just as selfish as my sharing it with a hungry stranger, language has become obsolete. How can a single concept cover such divergent motivations? More importantly, why is my satisfaction at seeing the stranger eat confused with my being selfish? Why can't altruism be like any other natural human tendency in that it yields pleasure? Many people love to spoil their family and friends, and the greatest joy we can give them is to just let them do it.

Looking back on how we got to this reversal of perspectives—from altruism as a sacrifice that is hard to explain to the modern notion of altruism as rooted in mammalian nurturance endowed with intrinsic rewards—I am struck by how many ideological and religious elements infused the debate, ranging from Price's conversion to Christianity, Huxley's preoccupation with original sin, Kropotkin's anarchism, and the curiously popular notion of altruism as either hypocritical or mistaken. Missing from most of this debate has been the view that humans and other mammals achieve altruism quite differently than, say, the social insects. Perhaps it is the common comparison of human altruism with that of ants and bees that has thrown us off. Insects lack empathy, whereas our brains are built to connect with others and experience their pain and pleasure. The end result is that altruism can be both genuine and satisfying at the same time. If Price pushed himself beyond this boundary, by giving up his health and wealth for fellow vagabonds whom he barely knew, his eventual despair is understandable. He underestimated the hedonic quality of altruism aimed at those we care about, and overestimated our capacity for generosity toward strangers. The second is seriously limited, whereas the first knows few bounds.

Chapter 3

BONOBOS IN THE FAMILY TREE

The best way to destroy an enemy is to make him a friend.

—Abraham Lincoln[1]

A visit to a forensic laboratory in Moscow confirmed my suspicion that being a close relative of the crown of creation won't get you any respect. The lab specialized in forensic sculpting, such as facial reconstructions from the skulls of unidentified murder victims. In a corner of the basement, my hosts revealed a rough-hewn face that they were trying to keep under wraps. I wasn't even allowed to take a picture. They had tried their hand at a Neanderthal skull, and the resulting bust so eerily resembled one of the most powerful Duma (parliament) members that they feared he would close down their institute if a picture ever saw the light of day.

We don't have a particularly high opinion of our next of kin and certainly don't want to look like them. Neanderthals are depicted as stooped retards running in and out of caves and dragging their wives behind them by their hair. The latest discovery, the small "hobbit" from the Indonesian island Flores, is considered microcephalic, perhaps even a "cretin." Scientists blame thyroid problems caused by lack of iodine, but most of us go by the dictionary definition of a cretin as a "stupid, obtuse, or mentally defective person." Never mind that sophisticated tools were found close to the Flores fossil.

Staging the unsightly fights they're known for, anthropologists are still pondering the Flores evidence, but the situation of Neanderthals is getting clearer by the day. The traditional stereotype that they are dumb brutes never made much sense, given that their brain size exceeded ours. The Duma member's resemblance rather tells us something about our own Neanderthal background. When early humans traveled out of Africa, they encountered close relatives who had already spent a quarter million years up north. These relatives were far better adapted to the freezing cold. Instead of us conquering them, as the story goes, we may have befriended the northerners. Men must have thought Neanderthal women were hot, women must have fancied Neanderthal men, and the other way around, because it is estimated that up to 4 percent of the DNA of non-African members of our species stems from Neanderthals. The crossbreeding probably boosted our immune system.

Our northern brethren buried their dead, were skilled toolmakers, kept fires going, and took care of the infirm just like early humans. The fossil record shows survival into adulthood of individuals afflicted with dwarfism, paralysis of the limbs, or the inability to chew. Going by exotic names such as Shanidar I, Romito 2, the Windover Boy, and the Old Man of La Chapelle-aux-Saints, our ancestors supported individuals who contributed little to society. Survival of the weak, the handicapped, the mentally retarded, and others who posed a burden is seen by paleontologists as a milestone in the evolution of compassion. This communitarian heritage is crucial in relation to this book's theme, since it suggests that morality predates current civilizations and religions by at least a hundred millennia.

This is not the only date that is being pushed back, though. It's a safe assumption that, ranging from the brewing of beer to expressions of art, things always happened earlier than thought. South African ocher pieces engraved with complex geometrical patterns are twice as old as the cave paintings in Lascaux, France. Even bipedal locomotion keeps getting pushed back—for example, by the discovery of footprints showing a fully upright gait twice as long ago than previously assumed.

Our first assumption is invariably that what we do and are proud of must be a recent development. Then we discover that Neanderthals did the same, and perhaps Australopithecines, too, until we go back in time all the way to the apes, which may in fact have been the first. Who says, for example, that the Stone Age started in our lineage? By means of archaeological techniques, a more than 4,000-year-old nut-cracking site with hammer and anvil stones was unearthed in Ivory Coast, but the kind of nuts found, the size of the tools (large and heavy), and the ecology (rainforest) hint at chimpanzee users rather than at human ones. Analysis of the excavation suggests that for thousands of years apes have been bringing durable stones, such as granite, from distant outcrops to smash hardy nuts in the forest. Today, the same tool technology is well known among West African chimpanzees.

The Long Farewell

There is only one date that, rather than being pushed back, has been creeping closer and closer. In the first half of the preceding century, evolutionary trees in textbooks still showed the human branch proudly growing on its own for 25 million years.

Our immediate family consists of the four great apes (chimpanzees, bonobos, gorillas, and orangutans) and the so-called lesser apes: gibbons and siamangs. This is a tiny family compared with the two hundred species of monkeys and prosimians in the primate order. With their tails and protruding snouts, monkeys are more distant from us than the apes are. The old tree, which put us far apart from all other primates, was not to last, however. Carl Linnaeus may already have foreseen this when he assigned humanity its own separate genus, *Homo*. The story goes that the Swedish taxonomist had his doubts about our special status but decided to avoid trouble with the Vatican. Three centuries later, the analysis of blood proteins and DNA offered a better way of comparing species than the anatomical comparisons used until then. The new data placed us apart from the monkeys, but smack in the middle of the apes. This was a shocker, but DNA is hard to argue with, because it avoids the problem of humans' picking and choosing traits they like to emphasize. We may think that walking on two legs is a big deal, yet in the larger scheme of nature it really isn't. Chickens do it, too. DNA comparisons circumvent human bias. In a DNA-based tree, humanity occupies just one tiny branch among many, having split off from the apes around six million years ago.

If crossbreeding toward the end of the road (such as with the Neanderthals) fueled our species' success, the same may have applied to the beginning. Human and ape DNAs show signs of early hybridiza-

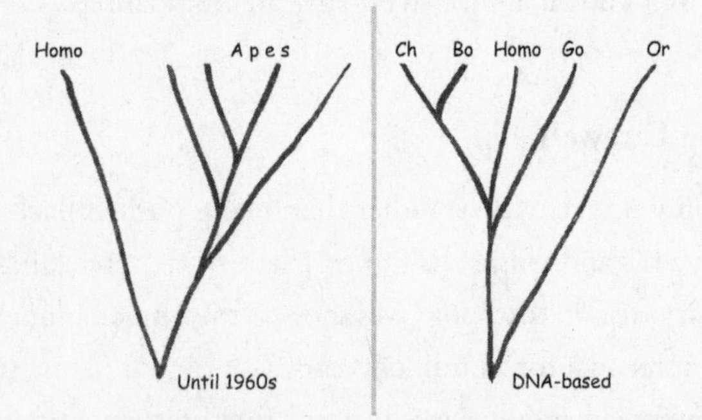

Until the 1960s, humans enjoyed their own branch on the evolutionary tree separate from the apes (left). DNA-based trees (right), however, place humans closer to chimpanzees (Ch) and bonobos (Bo) than to gorillas (Go) and orangutans (Or).

tion. After having split off, our ancestors probably kept returning to the apes in the same way known today of grizzlies and polar bears or wolves and coyotes. Some paleontologists are skeptical, calling it unlikely that our bipedal ancestors kept breeding for over a million years with apes that walked on all fours, but to my knowledge the way you walk says little about whether you can mate or not. It makes me think of an even more puzzling assertion from before we knew about human-Neanderthal hybridization, which was that we could safely rule out sex between those two hominins, since they obviously didn't speak the same language. I had to chuckle at this, thinking about when my French wife and I first met. Language is such a minor barrier.

The first to propose human descent from the apes was the French naturalist Jean-Baptiste Lamarck, in 1809. According to Lamarck's theory, acquired characteristics (such as the stretching of legs by wading birds) can be passed on to the next generation. Long before Darwin touched the subject, Lamarck envisioned human evolution from a quadrumanous (four-handed) primate:

> *If some race of quadrumanous animals, especially one of the most perfect of them, were to lose, by force of circumstances or some other cause, the habit of climbing in trees and grasping branches . . . and if the individuals of this race were forced for a series of generations to use their feet only for walking, and to give up using their hands like feet . . . these quadrumanous animals would at length be transformed into bimanous, and the thumbs of their feet would cease to be separated from the other digits.*[2]

Lamarck paid dearly for his audacity. He made himself so many enemies that he died in penury and was the subject of one of the most mocking and demeaning obituaries ever read before the Académie des Sciences.[3] Half a century later, human descent from apes was popularized by two champions of Darwinian evolution—the kind based on inherited traits—Thomas Henry Huxley, in England, and Ernst Haeckel, in Germany. Those two fought hard to get people to accept that we are modified apes, and they convinced at least the scientific community, in which this isn't a point of discussion anymore. Except, that is, when in 2009 Kent State University came out with a press release under the shocking headline "Man Did Not Evolve from Apes."

To understand this claim one needs to know that Kent State had been involved in the discovery of *Ardipithecus ramidus*, also known as "Ardi," a 4.4-million-year-old fossil from Ethiopia. This puts Ardi one million years closer to the long goodbye between humans and apes than previous fossils. One sign that Ardi was still rather apelike is her opposable big toe. She must have been a great climber, who slept in trees the way apes still do every night to avoid predators. Inevitably, creationists and intelligent-designers jumped on the misleading press release as a gift from God, while media outlets concluded that this must mean that apes descend from us. The confusion arose because a scientist on the Ardi team, despite being blessed with the bonobo-like name of Owen Lovejoy, could think only of chimps as a comparison. He concluded that Ardi's physique was too different to have come from a chimp-like forebear. But why take any living ape as starting point? The apes that are around today have had as much time to change as our own species has had since we split. People often think that the apes must have stood still while we evolved, but genetic data in fact suggest that chimpanzees changed *more* than we did. We simply don't know what our last common ancestor looked like. The rainforest doesn't permit fossilization—everything rots away before it gets to this point—which is why we lack early ape fossils. Nevertheless, we can be sure that our progenitor would fit the common definition of an ape: a large, tail-less, flat-chested primate with grasping feet. It remains perfectly acceptable, therefore, to say that we descend from apes, just not from any of the current ones.

Ardi's less protruding mouth and relatively small, blunt teeth clearly set her apart from the chimpanzee in which males are equipped with long, sharp canines. These "fangs" serve as deadly knives, capable of slashing open an enemy's face and skin. Wild chimps use this weaponry to lethal effect in territorial combat. In comparison, Ardi is thought to have been relatively peaceful, perhaps because of reduced male-to-male conflict. Lovejoy even proposed that Ardi and her contemporaries were monogamous and that this helped them suppress violence. But unless paleontologists come up with a male and female fossil with wedding rings on, the idea that Ardi was pair-bonded remains pure speculation. Moreover, there is no evidence that monogamy fosters peacefulness: the only pair-bonded primate we have in our immediate lineage (the gibbon) has formidable canine teeth.

What if we descend not from a blustering chimp-like ancestor but from a gentle, empathic bonobo-like ape? The bonobo's body proportions—its long legs and narrow shoulders—seem to perfectly fit the descriptions of Ardi, as do its relatively small canines. Why was the bonobo overlooked? What if the chimpanzee, instead of being an ancestral prototype, is in fact a violent outlier in an otherwise relatively peaceful lineage? Ardi is telling us something, and there may exist little agreement about what she is saying, but I hear a refreshing halt to the drums of war that have accompanied all previous scenarios.

The typecasting of our ancestors and relatives often has political overtones, which turned humorous in the hands of the comedian Stephen Colbert on *The Colbert Report.*[4] It's quite an experience to go on a show the main purpose of which is to make fun of you and your ideas. Colbert asked me how chimpanzees and bonobos stack up against each other. While I explained bonobo behavior, he kept pulling disgusted faces: they clearly were too peaceful and sexy for his taste ("What about regular, straight, as-God-intended-it sex?"). But he nodded with approval when I described chimpanzees, which perfectly fit his law-and-order character.

Genito-genital rubbing between bonobo females serves bonding and peacemaking. The two females press their vulvas and clitorises together and frantically rub them sideways against each other, while one female clings to the other almost like an infant. Facial expressions and loud squeals suggest orgasm.

In a world divided by chimpophiles and bonobophiles, we all had a good laugh when Stephen peeled his banana.

Bonobos, Left and Right

Imagine you're a writer, and you have decided to offer your readers a firsthand account of the politically correct primate, the idol of the left, known for its "gay" relations, female supremacy, and pacific lifestyle. Your focus is the bonobo: a close relation of the chimpanzee. You travel all the way to a place ironically called the Democratic Republic of the Congo (DRC) to see these darling apes frolic in their natural habitat, hoping to come back with new and exciting stories.

Alas, you barely get to see any bonobos. You watch a few of them quietly sitting in the trees, eating nuts. That's all. This is what happened to Ian Parker, who nevertheless managed to write thirteen pages of carefully crafted prose as a "far-flung correspondent" for *The New Yorker.* We learn about the "hot, soupy air," the rain storms, the mud streams, the sound of falling fruit shells, and his German host, whom he describes as rather cold and unsympathetic. Parker's main message could of course have been that fieldwork is no picnic, but instead he insisted that bonobos were not nearly as nice and erotic as people think. Given that this ape's reputation has been a thorn in the side of homophobes as well as Hobbesians, the right-wing media jumped with delight. The bonobo "myth" could finally be put to rest, and nature remain red in tooth and claw. The conservative commentator Dinesh D'Souza accused "liberals" of having fashioned the bonobo into their mascot, and he urged them to stick with the donkey.

This might all have been amusing if it hadn't been for the fact that these are not just political skirmishes. At issue is what we *know*. That bonobos can be aggressive is not in doubt. We know of fierce group attacks, mostly by females against males. Many such cases have been documented at zoos over the years, and have actually led to changes in how bonobos are being kept. Since separation of mothers and sons disrupts a protective bond, zoos are increasingly keeping them together. As I warned in my 1997 book *Bonobo: The Forgotten Ape*, "All animals are competitive by nature and cooperative only under specific circumstances."[5]

With regard to bonobo behavior in the field, there are few new discoveries. The DRC has only recently emerged from a bloody civil war that killed an estimated five million people—an atrocious situation that has not been conducive to primatological research. Knowledge about wild bonobos has been at a virtual standstill for over a decade. We have excellent field data from before that time, however. The most important observation, which has remained unchanged over the last three decades, is that there are no confirmed reports of lethal aggression among bonobos. For chimpanzees, in contrast, we have dozens of cases of adult males killing other males, of males killing infants, of females killing infants, and so on. This is in the wild. In captivity, I myself documented male chimpanzees brutally mutilating and castrating a political rival, which led to his death. There is absolutely no dearth of such information on chimpanzees, which contrasts sharply with the zero incidence in bonobos.

Reviewing the violence of chimpanzees in *Demonic Males*, Richard

Wrangham went on to draw the following comparison with bonobos: ". . . we can think of them as chimpanzees with a threefold path to peace. They have reduced the level of violence in relations between the sexes, in relations among males, and in relations between communities."[6] None of this is to say that bonobos live in a fairy tale. They engage in "sex for peace" precisely because they have plenty of conflicts. What would be the point of peacemaking if they lived in perfect harmony? Sexual conflict resolution typically occurs among females, but also among males, such as at the San Diego Zoo:

> *Vernon regularly chased Kalind into the dry moat. . . . After such incidents the two males had almost ten times as many intensive contacts as normal for them. Vernon would rub his scrotum against Kalind's buttocks, or Kalind would present his penis for masturbation.*[7]

The contrast with their fellow species is striking. Most observed chimp killings take place during territorial disputes, whereas bonobos engage in sex at their boundaries. They can be unfriendly to neighbors, but soon after a confrontation has begun, females have been seen rushing to the other side to copulate with males or mount other females. Since it is hard to have sex and wage war at the same time, the scene rapidly turns into socializing. It ends with adults from different groups grooming each other while their children play. These reports go back to 1990, and come mainly from Takayoshi Kano, the Japanese scientist who worked the longest with wild bonobos. Writing *Bonobo*, I interviewed fieldworkers, such as Kano and Parker's host, Gottfried Hohmann. When I asked the latter how his bonobos react to other groups, Hohmann replied, "It starts out very tense, with shouting and chasing, but then they settle down and there is female-female and male-female sex between members of the two communities. Grooming may occur, but remains tense and nervous."[8] This is not exactly the stuff associated with killer apes, although Hohmann did add that communities do not always mingle and that males from different groups don't groom each other.

At a sanctuary near Kinshasa, it was recently decided to merge two bonobo groups that had lived separately, just so as to create some activity. No one would ever dream of doing such a thing with chimpanzees, because the only possible outcome would be violence. It is well known at zoos that chimpanzee strangers need to be kept apart at all cost until they have become acquainted; otherwise one may be facing a bloodbath. The bonobos at the sanctuary, however, produced an orgy instead. They mixed freely, turning potential enemies into friends.

To top it all off, there are the observations by Isabel Behncke, a Chilean primatologist, who studies bonobo play behavior at Wamba, the site where Kano and other Japanese scientists have worked for decades. Isabel couldn't believe her eyes when she saw individuals of different groups play together. She recently showed me videos taken in dense forest of an adult male surrounded by juveniles from a neighboring group, who were poking him, climbing on top of him, and dangling around him. It was all in fun, without a grain of danger or hostility. She also showed a game between a male and a female from an outside group, in which the female followed the male and grasped his testicles while both of them ran around and around a tree, again without any obvious tension. A bit of a playful character herself, Isabel joked that this is where the expression "holding him by the balls" must derive from.[9]

Part of the confusion about aggressiveness of bonobos comes from their predatory behavior. While not on the scale of that of chimps, it is well developed. Bonobos kill small game, such as duikers (forest antelopes), squirrels, and immature monkeys, and sometimes hunt in groups. The problem is that this has little to do with aggressiveness. Already in the 1960s, Konrad Lorenz warned that a cat hissing at another cat is not the same as a cat stalking a mouse. The first expresses a mixture of fear and aggression, the second is motivated by hunger. We know now that the neural circuitry is different. This is why Lorenz defined aggression as within-species behavior, and why herbivores are not considered any less aggressive than carnivores—as anyone who has witnessed a stallion fight can attest.

Confusing predation with aggression is an old error that recalls the time that humans were seen as incorrigible murderers on the basis of signs that our ancestors ate meat. This "killer ape" notion gained such traction that the opening scene of Stanley Kubrick's movie *2001: A Space Odyssey* showed one hominin bludgeoning another with a zebra femur, after which the weapon, flung triumphantly into the air, turned into an orbiting spacecraft. A stirring image, but based on a single puncture wound in the fossilized skull of an ancestral infant, known as the Taung Child. Its discoverer had concluded that our ancestors must have been carnivorous cannibals, an idea that the journalist Robert Ardrey repackaged in *African Genesis* by saying that we are risen apes rather than fallen angels.[10] It is now considered likely, however, that the Taung Child had merely fallen prey to a leopard or eagle.

Glorification of violence stands in contrast to our coyness about sex, which has led scientists either to ignore it or to label it something else. In the same way that we prefer euphemisms—calling a toilet a "restroom," or the unintended exposure of a nipple a "wardrobe malfunction"—the literature customarily calls bonobos "very affectionate," while in fact referring to behavior that, if conducted in the human public sphere, would promptly get you arrested. Two females may be pressing their swollen genitals together, rapidly rubbing them laterally in a pattern known as genito-genital rubbing, but Hohmann, who has seen this pattern many times, wonders, "But does it have anything to do with sex? Probably not. Of course, they use the genitals, but is it erotic behavior or a greeting gesture that is completely detached from sexual behavior?"[11]

Fortunately, a U.S. court settled this monumental issue in the Paula

Jones case against President Bill Clinton. It clarified that the term "sex" includes any deliberate contact with the genitalia, anus, groin, breast, inner thigh, or buttocks. We may quibble with this definition (if someone deliberately sits on me, thus touching me with his or her buttocks, does this need to be sexual?), but let's focus on the genitals, which are obviously made for sex. When bonobos stimulate each other by grabbing testicles, fingering clitorises, or rubbing genitals together while squealing and showing other signs of apparent orgasm, any sex therapist will tell you that they are "doing it." I am thinking here of Susan Block, an American therapist who teaches "The Bonobo Way of Peace through Pleasure," which seems an appropriate slogan, given that, apart from our own species, no other animal is as much into sex as the bonobo.[12]

How dramatically bonobos differ from chimpanzees was highlighted by a recent experiment on cooperation. Brian Hare and coworkers presented apes with a platform that they could pull close by working together. When food was placed on it, the bonobos outperformed the chimpanzees. The presence of food normally induces rivalry, but the bonobos engaged in sexual contact, played together, and happily shared the food side by side. The chimpanzees, by contrast, had trouble overcoming their competitiveness. For two species to react so differently to the exact same setup leaves little doubt about a temperamental difference.

Another piece of evidence comes from a comparison of ape orphans by Vanessa Woods. Sadly, both chimps and bonobos are frequent victims of bushmeat hunting in Africa. Adults are typically sold as meat, whereas infants often end up in sanctuaries where humans lovingly care for them until they are strong enough to fend for themselves. Woods made a detailed comparison of infants of both species and found that bonobo infants engage in sexual contact at moments of excitement, such as when they are being fed, whereas chimp infants don't. The species difference therefore shows up extremely early in life.

In short, so long as we dare to call sex "sex" and focus on known levels of within-species (as opposed to between-species) violence, there remains strong support for the claim that bonobos are relatively peaceful and that sexual behavior serves nonreproductive functions, including greeting, conflict resolution, and food sharing. The occasional hyperbole (for example, "chimpanzees are from Mars, bonobos are from Venus") may be going too far, but no one would ever have heard of the species had it been described as merely affectionate. Whatever we find out in the years to come, with bonobo fieldworkers returning to Africa, a Hobbesian makeover of the species is not to be expected anytime soon. I just can't see this ape go from being a gentle, sexy primate to being a nasty, violent one. The only scientist who has extensively studied both chimpanzees and bonobos in the forest, the

Are humans fallen angels? In *The Garden*, Bosch painted Adam and Eve being brought together by a Jesus-like figure. Unusually, the first humans are depicted neither as eating any forbidden fruits nor as facing eviction. Absent the Fall, did a paradise of lust await them?

Japanese primatologist Takeshi Furuichi, said it best: "With bonobos everything is peaceful. When I see bonobos they seem to be enjoying their lives."[13]

Paradise of Lust

As a student, I visited a now defunct Dutch zoo that kept "pygmy chimpanzees," which is the old name for bonobos. It was the first time I'd seen the species. I was struck by the contrast in behavior, demeanor, and appearance with chimpanzees. Chimps are brawny bodybuilders, whereas these apes looked rather intellectual. With their thin necks and piano-player hands, they seemed to belong in the library rather than the gym. At the time, almost nothing was known about bonobos, and I decided on the spot that this had to change. I had been led to believe that bonobos were just a smaller version of the chimpanzee, which was patently wrong.

That day, I witnessed a minor squabble over a cardboard box, in which a male and a female ran around and pummeled each other, when all of a sudden their fight was over and they were making love! This seemed odd: I was used to chimpanzees, who don't switch so easily from anger to sex. I thought that it was a coincidence, or that I had missed something that would explain the change of heart, but it turned out that what I had seen was perfectly normal for these Kama Sutra primates. I learned this only years later, however, after I had begun working with them.

Even if becoming a primate sexologist was never my goal, it was an inevitable consequence. I have seen them do it in all positions one can imagine, and even in some that we find hard to imagine (such as upside down, hanging by their feet). The most significant point about bonobo sex is how utterly casual it is, and how well integrated with social life. This is not how most of us look at our love life, since we are full of hang-ups, obsessions, and inhibitions. Some people can't even do it with the lights on! This is why everyone winks at me when I say I work with bonobos, as if it must be a thrill, a forbidden pleasure that I get to enjoy. But the more one watches bonobos, the more sex begins to look like checking your email, blowing your nose, or saying hello. A routine activity. We use our hands in greetings, such as when we shake hands or pat each other on the back, while bonobos engage in "genital handshakes." Their sex is remarkably short, counted in seconds, not minutes. We associate intercourse with reproduction and desire, but in the bonobo it fulfills all sorts of needs. Gratification is by no means always the goal, and reproduction only one of its functions. This explains why all partner combinations engage in it.

In discussing this multifunctionality, it's hard not to notice that some people hate it and some love it. The hate relates to established views about the role of male hierarchies, territoriality, and violence in human evolution, which is no doubt why anthropologists keep ignoring the bonobo. They have no room for primate hippies in our past. Bonobophilia is not necessarily more rational, however. It often reflects wishful thinking that is based on an idealized image of our ancestors. After lectures on bonobos, I sometimes run into polyamorists, who feel they have much in common with the species, or people who tell me they dream of being more like bonobos. Others speculate that we must be their direct descendants, while implying that we should become matriarchal and throw off our sexual straitjackets.

The association of our ancestors with free love has a biblical undertone. Not that the Bible encourages promiscuity, but it tells us that before the Fall we didn't know any better. Other primates are sometimes viewed as living innocent lives in a pristine environment, the way we envision Eden. They are thought to know no sexual bounds. There is even an official theory about this by the French anthropologist Claude Lévi-Strauss, who proposed that human civilization started with the incest taboo. Before that time, we did it with everyone regardless of whether we were blood relatives or not. The incest taboo pushed us into a new realm: from the natural into the cultural. How off the mark was Lévi-Strauss! Suppression of inbreeding, as biologists call it, is well developed in all sorts of animals, from fruit flies and rodents to primates. It is close to a biological mandate for sexually reproducing species. In bonobos, father-daughter sex is prevented by females' leaving around puberty to join neighboring communities. And mother-son sex is wholly absent, despite the fact that sons stick around and often travel with their mothers. It is the only partner combination free of sex in bonobo society. And all of this sans taboos.

Starting with Rousseau's "noble savage," prehistory is often reconstructed from an insouciant perspective in which we all happily get along without a moment's worry about tomorrow. Margaret Mead seemed drawn to this view when she described Samoan love life, and a recent BBC documentary pasted this particular Western prejudice onto an "untouched" Amazonian tribe. Two anthropologists familiar with the Peruvian Matsigenka community claim that the entire documentary was fabricated. It started with the way members of the film crew entered the village. They veered off a well-traveled path so that they could film themselves hacking their way through the jungle to find these elusive people. The documentary is said to be full of egregious mistranslations that turn innocuous remarks ("you come from far away where the *gringos* live") into fierce statements ("we use arrows to kill outsiders"). When the old village chief wistfully says, "I will have sex another day," the translation reads, "I have sex every day."[14]

Imagination runs amok as soon as we think of human origins. We picture ancestors without culture, before language, with hardly any technology, and a minimum of sexual constraints. It sounds rather unrealistic, but two centuries before Rousseau, people were getting ready for this genre of origin fantasies, as is evident from the almost simultaneous appearance of the English humanist Thomas More's *Utopia* and Bosch's *Garden*. More's world included a welfare state, lack of private property, and euthanasia, but it lacked free love. In fact, premarital sex was punished in Utopia by a lifetime of celibacy. Bosch's fantasy, in contrast, showed us a happy mass of undressed men and women frolicking around in *The Garden*'s central panel, indulging both their palates and their genitals. What did the painter try to tell us? The traditional interpretation is that his triptych shows the corruption of innocence followed by hellish punishment in the right-hand panel. It seemed straightforward: sex is sin, and sinners belong in hell. If true, *The Garden*'s moral outlook doesn't differ much from *Utopia*'s. But we know now that Bosch's painting reveals its secrets only reluctantly. Together with Leonardo da Vinci's *Last Supper*, which dates from the same time, it is probably the most written-about work of art ever. It appears new to every successive generation—often revealing more about the generation's era than about the painting itself.

The paradise panel on the left shows God gently holding Eve's wrist with his left hand while blessing her union with Adam with his right hand. Adam stares at Eve with what has been described as sexual arousal. For this to be true, however, and speaking as a primatologist here, I would have preferred to see an erection. Yet, Adam's member appears as unanimated as a sleeping mouse (Google Earth lets anyone violate Adam's privacy). His facial expression, rather, seems astonished, as if no one had ever told him that he was going to meet a woman. The first couple meets in a most unusual setting, full of invented creatures and recently discovered animals (giraffes, porcupines). In the distance, we see a snake curling around a sort of nut tree, but the snake is in fact dropping out of the tree, and Adam and Eve don't eat any fruits. Indeed, *The Garden* shows us Eden with neither a fall nor an expulsion.

It took art historians a couple of centuries to figure out that the horizon in the middle panel is continuous with that of the left-hand side, suggesting that the central scene with over a thousand nudes engaged in erotic amusement must be taking place in paradise, too. Is this what might have become of humanity had we never been kicked out? Would resistance to temptation have been rewarded with sexual freedom? In the words of one art critic, the swarms of men riding donkeys, camels, and four-legged birds around pools with bathing ladies exhibit "a certain adolescent sexual curiosity."[15] This is precisely how bonobos have been described by me and others. The species seems an immature version of the chimpanzee in the same way that humans are considered a forever-young primate. *Neoteny*, or the retention of juvenile traits into adulthood, is considered the hallmark of our species. It is recognizable in our enduring playfulness, inquisitiveness, and creativity, and also in our imaginative sexuality. *The Garden* offers a perfect illustration, and since it abounds with all sorts of animals, I wager that Bosch, had he known bonobos, would not have hesitated to insert a few among the cavorting masses. They'd have fit in far better than chimps.

Some have related Bosch's intentions to the fourth-century Latin translation of the Bible, the Vulgate, which speaks of the *paradisum voluptatis*, or "the paradise of lust." Bosch undoubtedly knew this translation by Saint Jerome, his namesake, whom he admired so much that he painted him twice.[16] Early theologians were embarrassed by the Vulgate's reference to lust, pleasure, and enjoyment, but couldn't deny that God had created humans in two versions with complementary genitals. His well-known injunction to be fruitful and multiply could not possibly be carried out without sex and the gratification that comes with it.

Bosch may have taken all of this quite literally while adding some deliberate provocation. One popular speculation is that he was part of a heretical sect, the so-called Brethren and Sisters of the Free Spirit, which sought a return to humanity's original purity, including nudity and promiscuity. Trying to achieve the innocence of Adam before the Fall, members of this sect were known as Adamites. There is no corroborating evidence, however, that Bosch was in fact an Adamite; the sect had entirely disappeared from view by the time he was born. More likely, he was influenced by early humanism, which was considerably less sex-averse than the church. The so-called Prince of the Humanists, Erasmus of Rotterdam, even stayed in Den Bosch to study Latin, living just a few houses down from the Bosch residence in the same street. It is tempting to speculate that those two satirizing moralists got to know each other.

Erasmus was crystal clear on sexuality:

> *I have no patience with those who say that sexual excitement is shameful and that venereal stimuli have their origin not in nature, but in sin. Nothing is so far from the truth. As if marriage, whose function cannot be fulfilled without these incitements, did not rise above blame. In other living creatures, where do these incitements come from? From nature or from sin?*[17]

This is quite a statement for the sixteenth century! It was all part of the moral and religious debates raging during the Northern Renais-

The middle panel of *The Garden* teems with nude figures, birds, horses, and imaginary animals. Fruits are freely consumed by the masses. Here a goldfinch holds up a blackberry for people, who reenact the old Dutch children's game of "cake-biting." Others engage in amorous pursuits or dream away.

sance. Bosch's oeuvre offered penetrating commentary. The painter's popular image may be as the morbid painter of punishment ("the master of the monstrous . . . the discoverer of the unconscious," as the psychoanalyst Carl Jung called him), but it is good to realize that among the tormented figures in the right-hand panel of *The Garden* we don't recognize any of the lovebirds from the center. There are references to lust and sex, but most of the depicted vices relate to gambling, greed, gossiping, sloth, gluttony, pride, and so on. It is almost as if the painter is saying that, yes, the world is full of misery and sin, and sin will be punished, but don't look at carnal love as its source.

Sisterhood Is Powerful

Vernon, a male bonobo at the San Diego Zoo, ruled a small group that included one female, Loretta, who was his mate and friend, and a couple of juveniles. It was the only time I have seen a bonobo group run by a male. At the time, I thought this must be the norm: after all, male dominance is typical of most mammals, and male bonobos are definitely larger and more muscular than females. But Loretta was relatively young, and she was the only female. As soon as a second one was added, the power balance shifted.

The first thing Loretta and the other female did upon meeting was engage in sex with big grins on their faces, squealing loudly, leaving little doubt that apes know sexual pleasure. These lesbian encounters became more and more frequent, spelling the end of Vernon's rule. Months later, the typical scene at feeding time was both females having sex and sharing the food. The only way for Vernon to get any was to beg with outstretched hand. What a contrast with chimpanzee groups, in which every healthy male dominates every female!

Female dominance is also typical of wild bonobos, as Furuichi explains:

> *When females approached males who were feeding in a preferred position at a feeding site, males yielded their positions to late-arriving females. Furthermore, males usually waited at the periphery of the feeding site until females finished eating. When overt conflict occurred, allied females sometimes chased males, but males never formed aggressive alliances against females. Even the alpha male might retreat when approached by middle- or low-ranking females.*[18]

One way to look at this unusual society is that it evolved to keep the young safe. Male chimpanzees occasionally kill infants of their species, and humans are not much better. Abuse and death may occur in the family home, but also on a larger scale, as when King Herod "sent forth, and slew all the children that were in Bethlehem, and in all its borders from two years old and under" (Matthew 2:16). Nothing like this happens in bonobos, either on a small or on a large scale. The reason is that, first of all, being the dominant sex helps mothers defend their children. Second, rampant sex makes every adult male potentially the father of every youngster. Not that male bonobos know about paternity, but what could possibly be worse than killing your own progeny? Such behavior is sure to be selected against, which is why promiscuity protects the young. This is visible in the behavior of new bonobo mothers. Instead of staying away from large gatherings, as chimp mothers wisely do, they join their group right away after having given birth. Bonobo mothers act as if they have nothing to fear.

Against this background, the only report of fierce violence among wild bonobos makes sense. Hohmann and his wife, Barbara Fruth, witnessed a dark incident in Lomako Forest concerning a young male named Volker. Volker had the luck that his mother, Kamba, was alpha female of the Eyengo community in which he was born. Bonobo males hang on to their mother's apron strings. As soon as Volker got into trouble with other males or was chased by females, his mother would step in, giving him a break. While growing up, Volker steadily improved his position among the males, climbing the social ladder with maternal support. He also developed a close friendship with a particular female, named Amy. Soon after Amy had given birth to her first offspring, however, an unexpected episode happened while a large number of bonobos foraged in a garcinia tree loaded with thousands of shiny sweet fruits:

> *Volker jumps on a branch that holds Amy and her baby. For a second the female seems to lose balance but then maintains a firm grip and pushes Volker off the branch. The male jumps to the ground and is followed by a screaming Amy. The descent of Volker and Amy initiates a rush, as other adult females and males drop out of the tree and within seconds the forest transforms into a battleground. Details are camouflaged by the dense vegetation, but the frightening noise of screaming bonobos indicates that this is not a mock fight but a fierce struggle.*[19]

The coordinated attack by fifteen or more apes was entirely aimed at Volker, who ended up being dragged about. He was eventually found on the ground desperately clinging to a tree with both hands and feet, his face contorted into a panicky grimace. All bonobos were agitated, having their hair on end and vocalizing, while warning off human observers with alarm barks, as if they didn't want them to get close. The faces of the bonobos showed emotions that Hohmann and Fruth had never seen before. The most surprising was that Amy, who was rather low-ranking, could trigger such a mass assault, while Kamba stayed entirely out of it. Normally, Kamba would have been the first to defend her son, but when the fieldworkers went looking for her after the incident, they found her hiding high up in the canopy.

The fieldworkers think that Volker may have threatened Amy's infant. Had he tried to snatch it away, the way chimpanzee males sometimes do? If so, Volker misjudged the community's protectiveness. A male who tries to lay hands on an infant is apparently in for the most dreadful punishment. The sudden outbreak of violence suggests a deeper layer to bonobo society, one normally covered by its Woodstock façade. It resembles a moral code to defend the interests of the most vulnerable. If violated, the code is so massively enforced that even the highest echelons of society, such as the alpha female, won't dare go against it.

Bonobo solidarity is made possible by a habitat that permits more social cohesion than that of chimpanzees. In their quest for dispersed food, chimps need to split up into small parties or travel long distances alone. Bonobos are different. They stay together, wait for those who have slowed down, and join a chorus of "sunset calls" to bring the community together while building night nests high up in the trees. They obviously love company. Access to enormous fruiting trees as well as abundant nutritious herbs on the forest floor support their close-knit society, the core of which is a "secondary sisterhood." I call it "secondary" since female bonding isn't based on kinship. Being the migratory sex, females are largely unrelated within any given community.

Fascination with infants is, of course, typical of all mammals. But one striking encounter illustrates how much bonobos care. A coworker of mine, Amy Parish, had gotten to know several female bonobos at a zoo. These females embraced Amy as one of them, something they never did with me, since apes make precise gender distinctions among people. Loretta might sexually solicit me from across the moat (turning her genital swelling to me while peeking between her legs), but being a male I could never be part of the gynecocracy that is a bonobo society. Amy, in contrast, once even got good food tossed at her, as if she must have been hungry. When Amy visited her bonobo friends years later, she wanted to show them her newborn son. From behind glass, the oldest female briefly glanced at Amy's baby, but then ran into an adjacent room. She quickly returned to hold her own baby up against the glass so that the two infants could look into each other's eyes.

An Empathic Brain

In comparing bonobo society with ours, I see too many differences to fall for bonobo-inspired wishful thinking. I don't believe their free love would necessarily suit us. For one thing, evolution has given us our own way of protecting the young, which is exactly the opposite of the bonobo's. Instead of diluting paternity, humans fall in love and often commit to one person, at least one at a time. Through marriage and morally enforced fidelity, many societies try to clarify which males fathered which offspring. It is a highly imperfect attempt, with lots of philandering and uncertainties, but one that has taken us into quite a different direction. Universally, human males share resources with mothers and offspring, and help out with child care, which is virtually unheard of in bonobos and chimpanzees. Most importantly, male partners offer protection against other males.

In considering what we share with our ape relatives, the easiest comparison is in fact between male chimpanzees and men. Chimpanzee males hunt together, form coalitions against political rivals, and collectively defend a territory against hostile neighbors, yet at the same time they vie for status and compete for females. This tension between bonding and rivalry is very familiar to human males on sports teams and in corporations. Men intensely compete among themselves while still realizing that they need each other to prevent their team from going under. In *You Just Don't Understand*, the linguist Deborah Tannen reports how men use conflict to negotiate status, and actually enjoy sparring with friends. When things have gotten heated, they make up with a joke or apology. Businessmen, for example, will shout and bully at a meeting, only to take a restroom break during which they joke and laugh it all off.

The fuzzy line between conflict and cooperation is not always understood by women (for whom a friend and a rival are totally different things), but it is second nature to me since I grew up in a family of six boys and no girls. In fact, my interest in how chimpanzees reconcile after fights came about partly because I refused to view aggression as inherently evil, which was the prevailing opinion when I began my studies. Aggressive behavior was even labeled "asocial." I failed to follow this. I saw scuffles and fights as a way of negotiating relationships, and would call them destructive only if inhibitions were lacking or if no one attempted a repair afterwards. Chimpanzee males get along most of the time and are indeed much better than females in reducing tensions through a long grooming session with their greatest rival. Holding grudges is not a male thing.

But I also see similarities with bonobos, especially when it comes to empathy and the social functions of sex. Not that humans use sex as easily and publicly as bonobos, but within the human family, sex serves as a social glue similar to the way it smooths relations among bonobos. I consider bonobos highly empathic, more so than chimpanzees. As soon as one bonobo has even the smallest injury, he or she will be surrounded by others who come to inspect, lick, or groom. Robert Yerkes described in *Almost Human* how his bonobo took care of a gravely ill companion, saying that if he were to give a full description, he would probably be accused of "idealizing an ape."[20]

It is only recently that we have learned how the brain of bonobos reflects this sensitivity. The first hint came from a special type of neuron, known as a spindle cell, thought to be involved in self-awareness, empathy, sense of humor, self-control, and other human fortes. Initially, these neurons were known only in humans, but following the usual pattern in science, they were subsequently also discovered in brains of apes, including bonobos.[21] Then came a study that compared specific brain areas in chimpanzees and in bonobos. Areas involved in the perception of another's distress, such as the amygdala and anterior insula, are enlarged in the bonobo. Its brain also contains well-developed pathways to control aggressive impulses. Reporting these neurological differences, James Rilling and co-workers concluded that bonobos have empathic brains:

> *We suggest that this neural system not only supports increased empathic sensitivity in bonobos, but also behaviors like sex and play that serve to dissipate tension, thereby limiting distress and anxiety to levels conducive with prosocial behavior.*[22]

None of this was known when I first encountered bonobos, yet it confirms that I was right to think they were different. The French call them "Left Bank chimpanzees," both for their alternative lifestyle and because they live on the south bank of the westward-streaming Congo River. This mighty river permanently cuts them off from chimpanzees and gorillas to the north. They share an ancestor with both of those apes, though, of which the one shared with chimps lived less than two million years ago. The $64,000 question is whether this ancestor was bonobo-like or chimp-like. In other words, which of the two apes is the more original type, closest in appearance and behavior to the sort of ape we derive from? For the moment, the safest bet is that chimps and bonobos are equally close to us, since they split from each other well after we split off from their ancestor. A common estimate is that we share 98.8 percent of our DNA with them, although other calculations put the number at "only" 95 percent.

The recent publication of the bonobo genome confirms that we humans share genes with bonobos that we don't share with chimps, but we also share genes with chimps that we don't share with bonobos.[23] While awaiting more precise DNA comparisons, it is clear that the argument that only chimps matter for the story of human evolution has now lost its footing. The bonobo is exactly equally relevant. Our species shares a mosaic of characteristics with both apes, or as I have said before, we are "bipolar apes." On our good days, we are as nice as bonobos can be, while on our bad days, we are as domineering and violent as chimps can be. How our common ancestor behaved remains unknown, but bonobos offer essential insights. They never left the humid rainforest, whereas chimps radiated out to half-open woodland and our own lineage left the forest altogether. So, bonobos may have encountered the least reasons for evolutionary change, and retained more original traits. The American anatomist Harold Coolidge suggested the same when, in 1933, he concluded from his necropsies that bonobos resemble "more closely to the common ancestor of chimpanzees and man than does any living chimpanzee."[24]

Chapter 4

IS GOD DEAD OR JUST IN A COMA?

It is useless to attempt to reason a man out of a thing he was never reasoned into.

—Jonathan Swift[1]

One quiet Sunday morning, I stroll down the driveway of my home in Stone Mountain, Georgia, to pick up the newspaper. As I arrive at the bottom—we live on a hill—a Cadillac drives up the street and stops right before me. A big man in a suit steps out, sticking out his hand. A firm handshake follows, during which I hear him proclaim in a booming, almost happy voice, "I'm looking for lost souls!" Apart from perhaps being overly trusting, I am rather slow and had no idea what he was talking about. I turned around to look behind me, thinking that perhaps he had lost his dog, then corrected myself and mumbled something like, "I'm not very religious."

This was of course a lie, because I am not religious at all. The man, a pastor, was taken aback, probably more by my accent than by my answer. He must have realized that converting a European to his brand of religion was going to be a challenge, so he walked back to his car, but not without handing me a business card in case I'd change my mind. A day that had begun so promisingly now left me feeling like I might go straight to hell.

I was raised Catholic. Not just a little bit Catholic, like my wife, Catherine. When she was young, many Catholics in France already barely went to church, except for the big three: baptism, marriage, and funeral. And only the middle one was by choice. By contrast, in the southern Netherlands—known as "below the rivers"—Catholicism was important during my youth. It defined us, setting us apart from the above-the-rivers Protestants. Every Sunday morning, we went to church in our best clothes, we received catechism at school, we sang, prayed, and confessed, and a vicar or bishop was present at every official occasion to dispense holy water (which we children happily imitated at home with a toilet brush). We were Catholic through and through.

But I am not anymore. In my interactions with religious and nonreligious people alike, I now draw a sharp line, based not on what exactly they believe but on their level of dogmatism. I consider dogmatism a far greater threat than religion per se. I am particularly curious why anyone would drop religion while retaining the blinkers sometimes associated with it. Why are the "neo-atheists" of today so obsessed with God's nonexistence that they go on media rampages, wear T-shirts proclaiming their absence of belief, or call for a militant atheism?[2] What does atheism have to offer that's worth fighting for?

As one philosopher put it, being a militant atheist is like "sleeping furiously."[3]

Losing My Religion

I was too restless as a boy to sit through an entire mass. It was akin to aversion training. I looked at it like a puppet show with a totally predictable story line. The only aspect I really liked was the music. I still love masses, passions, requiems, and cantatas and don't really understand why Johann Sebastian Bach ever wrote his secular cantatas, which are so obviously inferior. But other than developing an appreciation of the majestic church music of Bach, Mozart, Haydn, and others, for which I remain eternally grateful, I never felt any attraction to religion and never talked to God or felt a special relationship. After I left home for the university, at the age of seventeen, I quickly lost any remnant of religiosity. No more church for me. It was hardly a conscious decision, certainly not one I recall agonizing over. I was surrounded by other ex-Catholics, but we rarely addressed religious topics except to make fun of popes, priests, processions, and the like. It was only when I moved to a northern city that I noticed the tortuous relationship some people develop with religion.

Much of postwar Dutch literature is written by ex-Protestants bitter about their severe upbringing. "Whatever is not commanded is forbidden" was the rule of the Reformed Church. Its insistence on frugality, black dress code, continuous fight against temptations of the flesh, frequent scripture readings at the family table, and its punitive God—all contributed greatly to Dutch literature. I have tried to read these books, but have never gotten very far: too depressing! The church community kept a close eye on everyone and was quick to accuse. I have heard shocking real-life accounts of weddings at which the bride and groom left in tears after a sermon about the punishment awaiting sinners. Even at funerals, fire and brimstone might be directed at the deceased in his grave so that his widow and everybody else knew exactly where he'd be going. Uplifting stuff.

In contrast, if the local priest visited our home, he could count on a cigar and a glass of *jenever* (a sort of gin)—everyone knew that the clergy enjoyed the good life. Religion did come with restrictions, especially reproductive ones (contraception being wrong), but hell was mentioned far less than heaven. Southerners pride themselves on their bon vivant attitude to life, claiming that there's nothing wrong with a bit of enjoyment. From the northern perspective, we must have looked positively immoral, with beer, sex, dancing, and good food being part of life. This explains a story I heard once from an Indian Hindu who married a Dutch Calvinist woman from the north. Although the woman's parents didn't have the faintest idea what a Hindu was, they were relieved that their new son-in-law was at least not Catholic. For them, belief in multiple deities was secondary to the heretic and sinful ways of their next-door religion.

The southern attitude is recognizable in Pieter Brueghel's[4] and Bosch's paintings, some of which bring to mind Carnival, the beginning of Lent. Carnival is big in Den Bosch, when the city is known as *Oeteldonk*, and also celebrated in nearby Catholic Germany, in cities like Cologne and Aachen, where Bosch's family came from (his father's name, "van Aken," referred to the latter city). Bosch must have been well versed in the zany Carnival atmosphere, and its suspension of class distinctions behind anonymous masks. Just like Mardi Gras in New Orleans, Carnival is deep down a giant party of role reversal and social freedom. *The Garden of Earthly Delights* achieves the same by depicting everyone in his or her birthday suit. I am convinced that Bosch intended this as a sign of liberty rather than the debauchery some have read into it.

Possibly, the religion one leaves behind carries over into the sort of atheism one embraces. If religion has little grip on one's life, apostasy is no big deal and there will be few lingering effects. Hence the general apathy of my generation of ex-Catholics, which grew up with criticism of the Vatican by our parents' generation in a culture that diluted religious dogma with an appreciation of life's pleasures. Culture matters, because Catholics who grew up in papist enclaves above the rivers tell me that their upbringing was as strict as that of the Reformed households around them. Religion and culture interact to such a degree that a Catholic from France is really not the same as one from the southern

Netherlands, who in turn is not the same as one from Mexico. Crawling on bleeding knees up the steps of the cathedral to ask the Virgin of Guadalupe for forgiveness is not something any of us would consider. I have also heard American Catholics emphasize guilt in ways that I absolutely can't relate to. It is therefore as much for cultural as religious reasons that southern ex-Catholics look back with so much less bitterness at their religious background than northern ex-Protestants.

Egbert Ribberink and Dick Houtman, two Dutch sociologists, who classify themselves, respectively, as "too much of a believer to be an atheist" and "too much of a nonbeliever to be an atheist," distinguish two kinds of atheists. Those in one group are uninterested in exploring their outlook and even less in defending it. These atheists think that both faith and its absence are private matters. They respect everyone's choice, and feel no need to bother others with theirs. Those in the other group are vehemently opposed to religion and resent its privileges in society. These atheists don't think that disbelief should be kept locked up in the closet. They speak of "coming out," a terminology borrowed from the gay movement, as if their nonreligiousness was a forbidden secret that they now want to share with the world. The difference between the two kinds boils down to the privacy of their outlook.

I like this analysis better than the usual approach to secularization, which just counts how many people believe and how many don't. It may one day help to test my thesis that activist atheism reflects trauma. The stricter one's religious background, the greater the need to go against it and to replace old securities with new ones.

Serial Dogmatism

Religion looms as large as an elephant in the United States, to the point that being nonreligious is about the biggest handicap a politician running for office can have, bigger than being gay, unmarried, thrice married, or black. This is upsetting, of course, and explains why atheists have become so vocal in demanding their place at the table. They prod the elephant to see whether they can get it to make some room. But the elephant also defines them, because what would be the point of atheism in the absence of religion?

As if eager to provide comic relief from this mismatched battle, American television occasionally summarizes it in its own you-can't-make-this-stuff-up way. *The O'Reilly Factor* on Fox News invited David Silverman, president of the American Atheist Group, to discuss billboards proclaiming religion a "scam." Throughout the interview, Silverman kept up a congenial face, claiming that there was absolutely no reason to be troubled, since all that his billboards do is tell the truth: "Everybody *knows* religion is a scam!" Bill O'Reilly, a Catholic, expressed his disagreement and clarified why religion is *not* a scam "Tide goes in, tide goes out. Never a miscommunication. You can't explain that." This was the first time I had heard the tides being used as proof of God. It looked like a comedy sketch with one smiling actor telling believers that they are too stupid to see that religion is a fraud, but that it would be silly for them to take offense, while the other proposes the rise and fall of the oceans as evidence for a supernatural power, as if gravity and planetary rotation can't handle the job.[5]

All I get out of such exchanges is the confirmation that believers will say anything to defend their faith and that some atheists have turned evangelical. Nothing new about the first, but atheists' zeal keeps surprising me. Why "sleep furiously" unless there are inner demons to be kept at bay? In the same way that firefighters are sometimes stealth arsonists and homophobes closet homosexuals, do some atheists secretly long for the certitude of religion? Take Christopher Hitchens, the late British author of *God Is Not Great*. Hitchens was outraged by the dogmatism of religion, yet he himself had moved from Marxism (he was a Trotskyist) to Greek Orthodox Christianity, then to American Neo-Conservatism, followed by an "antitheist" stance that blamed all of the world's troubles on religion.[6] Hitchens thus swung from the left to the right, from anti–Vietnam War to cheerleader of the Iraq War, and from pro to contra God. He ended up favoring Dick Cheney over Mother Teresa.

Some people crave dogma, yet have trouble deciding on its contents. They become serial dogmatists. Hitchens admitted, "There are days when I miss my old convictions as if they were an amputated limb,"[7] thus implying that he had entered a new life stage marked by doubt and reflection. Yet, all he seemed to have done was sprout a fresh dogmatic limb.

Dogmatists have one advantage: they are poor listeners. This ensures sparkling conversations when different kinds of them get together the way male birds gather at "leks" to display splendid plumage for visiting females. It almost makes one believe in the "argumentative theory," according to which human reasoning didn't evolve for the sake of truth, but rather to shine in discussion. Universities everywhere have set up crowd-pleasing debates between religious and antireligious intellectual "giants." One such debate took place in 2009 at a large science festival in Puebla, Mexico. My own contribution concerned a different, more scientific session, but I sat in the audience of four thousand when we were being warmed up for the ultimate war of words. Asked whether they believed in God, about 90 percent of the people raised their hand in affirmation. The debate itself was set up in a distinctly unintellectual fashion. The stage showed a boxing ring (ropes around poles, red boxing gloves dangling in the corner), and the speakers walked one by one onto stage to martial music. They were the usual suspects. Apart from Hitchens, we got Dinesh D'Souza, Sam Harris, the philosopher Dan Dennett, and Rabbi Shmuley Boteach.

I would be surprised if a single member of the audience changed his or her mind as a result of the debate, either from believer to nonbeliever or the other way around. We learned that religion is the source of all evil and inferior to science as a guide to reality, but also that without religion there would be no morality and no hope for those who fear death. Without God, moral rules are "nothing but euphemisms for personal taste," exclaimed the rabbi, waving his hands above his head as if throwing pizza dough. Others spoke in a humorless, almost menacing tone, as if anyone who'd ignore their message would inevitably get into trouble. God isn't a fun topic.

The circus-like atmosphere left me with my original question about evangelical atheists. It's easy to see why religions try to recruit believers. They are large organizations with monetary interests that do better, the more people join them. They erect cathedrals, like the one I visited in Puebla, and chapels like the Capilla del Rosario with its 23½-carat gilded stucco. I've never seen such a blindingly ornate interior, probably paid for by generations of poor Mexican farmers. But why would atheists turn messianic? And why would they play off one religion against another? Harris, for example, biliously goes after the "low-hanging fruit" of Islam, singling it out as the great enemy of the West.[8] Throw in a few pictures of burqas, mention infibulation, and who will argue with your revulsion of religion? I am as sickened as the next person, but if Harris's quest is to show that religion fails to promote morality, why pick on Islam? Isn't genital mutilation common in the United States, too, where newborn males are routinely circumcised without their consent? We surely don't need to go all the way to Afghanistan to find valleys in the moral landscape.

If some religions are worse than others, then some must be better. I'd love to hear the atheist perspective on what makes for a good religion, or the reason why different religions support different moralities. Could it be that religion and culture interact to the point that there is no universal morality? Instead of pondering such problems, audiences are stirred up to abhor practices alien to them, which is about as easy as making them squirm at a chain-saw murder.

Then there is the persistent myth that science trumps religion in every possible way, and that science distracts from religion, and vice versa, as in a zero-sum game. This approach goes back to nineteenth-century American polemists, who famously declared that if it were up to religion, we'd still believe in a flat earth.[9] This was pure propaganda, however. Speculation about our planet's roundness began with Aristotle and other ancient Greeks, and every major scholar during the so-called Dark Ages was fully aware of it. Dante's *Divine Comedy* portrays the earth as a sphere, and the exterior panel of Bosch's *Garden* triptych takes an in-between approach by showing a flat earth floating in a transparent ball surrounded by a black cosmos. When it comes to evolution, too, there is a tendency to point at religion as a solid opponent while ignoring that the Roman Catholic Church never formally condemned Darwin's theory or put his works on the Index (the list of forbidden books). The Vatican has endorsed evolution as a valid theory compatible with the Christian faith. Admittedly, its endorsement came a bit late, but it is good to realize that resistance to evolution is almost entirely restricted to evangelical Protestants in the American South and Midwest.

The connection between science and religion has always been complex, including both conflict, mutual respect, and the church's patronage of the sciences. The first copiers of books on which science came to rely were rabbis and monks, and the first universities grew out of cathedral and monastic schools. The papacy actively promoted the establishment and proliferation of universities. At one of the first ones, in Paris, students cut their hair in tonsure to show allegiance to the church, and the oldest document in the archives of Oxford University is its Award of the Papal Legate of 1214. Given this intertwinement, most historians stress dialogue or even integration between science and religion.

Neo-atheists keep pitting the two against each other, however. Their audiences pee in their pants with delight when the flat-earth canard gets trotted out. This is not to say, however, that religious narratives are much better. They, too, play fast and loose with the facts.

In Puebla, D'Souza featured near-death experiences as scientific proof of the afterlife. After a brush with death, some patients report having floated outside of their bodies or having entered a tunnel of light. This surely seems bizarre, but D'Souza failed to bring up new neuroscience of a small brain area known as the temporo-parietal junction (TPJ). This area gathers information from many senses (visual, tactile, and vestibular) to construct a single image of our body and its place in the environment. Normally, this image is nicely coherent across all senses, so that we know who and where we are. The body image is disturbed, however, as soon as the TPJ is damaged or stimulated with electrodes. Scientists can deliberately make people feel that they are hovering above their own body or looking down on it, or have them perceive a copy of themselves sitting next to them, like a shadow ("I looked younger and fresher than I do now. My double smiled at me in

Bosch's *Ascent of the Blessed* (part of his *Visions of the Hereafter*) depicts the tunnel of light associated with near-death experiences that have inspired mythology and religion since the dawn of humanity.

a friendly way").[10] Together with the hallucinogenic qualities of anesthetic drugs and the effects of oxygen depletion on the brain, science is getting close to a materialist explanation of near-death experiences.

Rabbi Boteach, too, relied on questionable evidence to champion religion. He explained that many human families take care of Down syndrome children, which they obviously would never do without religion. They'd simply get rid of "defective" offspring, he said. The problem with this assertion, as mentioned in the preceding chapter, is that archaeological data tell quite a different story. Our lineage is equipped with such powerful nurturing instincts that offspring are not easily neglected or abandoned, no matter their condition. I'm not saying it never happens, but long before any of the current religions, Neanderthals and early humans took care of the handicapped. This is also true for our primate kin. There are many examples, but I'll limit myself to two that I know firsthand.

Azalea was a trisomic rhesus monkey; she had three copies of one chromosome, just as in human Down syndrome.[11] Another similarity was that she was born to a female beyond the age at which macaques normally conceive. Growing up in a large zoo troop, Azalea was seriously retarded in both motor development and social skills. She'd make the most incomprehensible blunders, such as threatening the alpha male. Rhesus monkeys are quick to punish anyone who breaks the rules, but Azalea got away with almost anything, as if the other monkeys realized that nothing they did would change her ineptness. We human observers were fond of Azalea for her sweet character, and it seemed that the rest of the troop was, too. She died naturally at three years of age.

Then there was Mozu, a Japanese macaque in the Japanese Alps, in Jigokudani. Mozu could barely walk, and certainly not climb, as she congenitally lacked both hands and feet. During the winter, which is severe in this area, she was forced to plow through the snow while her troopmates jumped from branch to branch. A frequent star of

Japanese nature documentaries, Mozu was fully accepted by the other monkeys to the point that she lived a long life and raised no fewer than five offspring. I saw her high up in the mountains and noticed that she spent most of her time with the other monkeys, away from people, so that the occasional handouts of food from tourists can't account for her survival. Even though there is no record of other monkeys' actively assisting her, her story goes to show that the unfit can thrive and reproduce in primate societies. Similarly, human life before religion was not necessarily dog-eat-dog. Instead of making us do things we normally wouldn't, religion may render its chief contribution by endorsing and promoting certain natural tendencies. This is obviously a much more modest contribution than what the rabbi had in mind.

Dogmatists pound their drums so hard that they can't hear one another. Their audiences, on the other hand, are unaware of the traveling dog and pony shows featuring the same adversaries over and over, who simulate surprise and "gotcha" moments. The only voice of reason in Puebla was that of Dan Dennett, who spoke about religion not as something hateful but rather as a phenomenon that begs investigation as part of human society, human nature even. Clearly, religion is

Azalea was a trisomic rhesus monkey, a condition similar to human Down syndrome. She was more dependent than other juveniles. Here she is being held like an infant by her older sister at an age where this isn't normal anymore. Despite her mental handicap, she was remarkably well integrated and accepted.

man-made, so the question is what good does it do for us. Are we born to believe and, if so, why? Dennett is not as sure as the neo-atheists with whom he is often lumped together that religion is irrational or that the world would be a better place if its demise were hastened, noting, "I am still agnostic about that."[12]

Droppings in a Cuckoo Clock

My first experience with proselytizing was almost too comical to be true. At the time, I had a room on the fourth floor of a university dormitory, in Groningen. One morning, I heard a knock on my door, and two young American Mormons in jackets and ties stood in front of me. Curious to hear about their faith, I invited them in. They proceeded to set up an easel and a board on which they pasted felt figures and text labels to explain the story of an ordinarily named American, who had seen the Lord in a pillar of light. Later, he was led by an angel to holy texts on golden plates.

All of this happened just over a century ago. I listened to their incredible tale and was just about to ask how this Joseph Smith convinced others of his special encounter, when we were rudely interrupted. I often left a window open to let Tjan, my pet jackdaw, fly in and out. He was free outside, but would come in before dark to be fed and locked up for the night. While the two young men patiently recounted God's appearance in a cave, Tjan sailed into the room looking for a landing spot. He went for the highest point, which was the head of one of the Mormons standing in front of his board. A large black bird landing on him was the last thing he'd expected. I saw the panic on his face and quickly tried to reassure him that this was just Tjan, a bird with a name, who wouldn't hurt anyone. I have never seen two people pack up so quickly: they were gone in no time, out the door, running for the elevator. While they collected their things, I heard them talk of the "devil."

My having spoiled Tjan as a baby with the fattest earthworms I could dig up made him extra large for his species. He was such a curious and intelligent character, who'd fly above me on strolls through the park. But of course he was black, noisy, and crow-like, reminding the Mormons of a creature that might steal their souls. As a result, they never got to answer my question, nor did they have a chance to explain how Smith translated the golden plates engraved with "reformed Egyptian" by gazing at "peep stones" placed in the bottom of his hat. Smith was smart enough to sympathize with his skeptics: "If I had not experienced what I have, I couldn't have believed it myself."[13]

So, why did people believe him? Smith met with a great deal of derision and hostility (he was killed by a lynch mob at the age of thirty-eight), but the Church of Jesus Christ of Latter-day Saints now counts 14 million followers. It is obvious that believers are not looking for evidence, because the only item that might have been helpful, the set of golden plates, had to be returned to the angel. People simply believe because they *want* to. This applies to all religions. Faith is driven by attraction to certain persons, stories, rituals, and values. It fulfills emotional needs, such as the need for security and authority and the desire to belong. Theology is secondary and evidence tertiary. I agree that what the faithful are asked to believe can be rather preposterous, but atheists surely won't succeed in talking people out of their faith by mocking the veracity of their holy books or by comparing their God with the Flying Spaghetti Monster. The specific contents of belief are hardly at issue if the overarching goal is a sense of social and moral communion. To borrow from a title by the novelist Amy Tan, to criticize faith is like trying to save fish from drowning. There's no point in catching believers out of the lake to tell them what is best for them while putting them out on the bank, where they flop around until they expire. They were in the lake for a reason.

Accepting that faith is driven by values and desires makes at once for a great contrast with science, but also exposes common ground, since science is less fact-driven than is widely assumed. Don't get me wrong, science produces great results. It has no competition when it comes to understanding physical reality, but science is also often, like religion, based on what we *want* to believe. Scientists are human, and humans are driven by what psychologists call "confirmation biases" (we love evidence that supports our view) and "disconfirmation biases" (we disparage evidence that undermines our view). That scientists systematically resist new discoveries was already the topic of a 1961 article in the illustrious pages of *Science*, which added the mischievous subheading "This source of resistance has yet to be given the scrutiny accorded religious and ideological sources."[14]

A good example is taste aversion. We remember food that has poisoned us so well that we gag at the thought of it. This reaction is of great survival value, yet it violated behaviorist dogma. Founded by B. F. Skinner, behaviorism claims that all behavior is shaped by reward and punishment, which works better the shorter the time interval between the act and its consequences. So, when the American psychologist John Garcia reported that rats avoid poisoned foods after just a single bad experience, even if the nausea sets in only hours later, no one believed him. Leading scientists made sure his study didn't appear in any mainstream journal. The author kept getting rejections, the most infamous one contending that his findings were no more likely than finding bird shit in a cuckoo clock. The "Garcia effect" is now well established, but the early reaction illustrates how much scientists hate the unexpected.

My own story concerns the discovery, in the mid-1970s, that chimpanzees make up after fights by kissing and embracing their opponents. Reconciliation behavior has now been demonstrated in many primates, but when one of my students needed to defend a study of this behavior before a committee of psychologists, she got an earful. We had naïvely assumed that these psychologists, who knew only rats, would have no opinion about primates, yet they were adamant that reconciliation in animals was out of the question. It didn't fit their thinking, which excluded emotions, social relationships, and everything else that makes animals interesting. I tried to change their minds by inviting them to the zoo where I worked so that they could see for themselves what chimpanzees do after fights. To this proposal, however, they replied bafflingly, "What good would it do to see the actual animals? It will be easier for us to stay objective without this influence."

It is said that the ancient king of Sardis complained that "men's ears are less credulous than their eyes." Only here it was reversed: these scientists feared that their eyes might tell them something they didn't want to hear. Like the rest of humanity, scientists apply flight-or-fight responses to data: they go for the familiar and avoid the unfamiliar. I have to think of this each time I hear neo-atheists claim that their God denial makes them smarter than believers and rational like scientists. They like to present themselves as emotion-free, "just the facts, ma'am" kind of thinkers. In a column in *USA Today*, Jerry Coyne, a fellow biologist and self-declared "gnu-atheist" (yes, "gnu" as in "wildebeest," which is Dutch for "wild beast"), called faith and science utterly incompatible "for precisely the same reason that irrationality and rationality are." He then proceeded to draw little aureoles around the heads of scientists:

> *Science operates by using evidence and reason. Doubt is prized, authority rejected. No finding is deemed "true"—a notion that's always provisional—unless it's repeated and verified by others. We scientists are always asking ourselves, "How can I find out whether I'm wrong?"*[15]

Oh, how I wish I had colleagues like Coyne! Having spent all my life among academics, I can tell you that hearing how wrong they are is about as high on their priority list as finding a cockroach in their coffee. The typical scientist has made an interesting discovery early on in his or her career, followed by a lifetime of making sure that everyone else admires his or her contribution and that no one questions it. There is no poorer company than an aging scientist who has failed to achieve these objectives. Academics have petty jealousies, cling to their views long after they have become obsolete, and are upset every time something new comes along that they failed to anticipate. Original ideas invite ridicule, or are rejected as ill informed. As the neuroscience pioneer Michael Gazzaniga complained in a recent interview,

> *There is a profound inhibitory effect on new ideas by people and ideas that "got there first," telling their story over and over while new observations struggle up from the bottom. The old line that human knowledge advances one funeral at a time seems to be so true!*[16]

This is more like the scientists I know. Authority outweighs evidence, at least for as long as the authority lives. There is no lack of historical examples, such as resistance to the wave theory of light, to Pasteur's discovery of fermentation, to continental drift, and to Röntgen's announcement of X-rays, which was initially declared a hoax. Resistance to change is also visible when science continues to cling to unsupported paradigms, such as the Rorschach inkblot test, or keeps touting the selfishness of organisms despite contrary evidence. Scientists praise the "plausibility" and "beauty" of theories, making value judgments on the basis of how they think things work, or ought to work. Science is in fact so value laden that Albert Einstein denied that all we do is observe and measure, saying that what we think exists is a product almost as much of theory as of observation. When theories change, observations follow suit.[17]

If faith makes people buy an entire package of myths and values without asking too many questions, scientists are only slightly better. We also buy into a certain outlook without critically weighing each and every underlying assumption and often turn a deaf ear to evidence that doesn't fit. We may even, like the psychologists on my student's committee, deliberately turn down a chance to get enlightened. But even if scientists are hardly more rational than believers, and even if the entire notion of unsentimental rationality is based on a giant misunderstanding (we cannot even *think* without emotions), there is one major difference between science and religion. This difference resides not in the individual practitioners but in their culture. Science is a collective enterprise with rules of engagement that allow the whole to make progress even if its parts drag their feet.

Darwinists Deserve a Darwin Award

What science does best is to incite competition among ideas. Science instigates a sort of natural selection, so that only the most viable ideas survive and reproduce. As an example, let's say that I believe that life is passed on through little homunculi inside sperm. You, in contrast, believe it's done by mixing the traits of both parents. Along comes an obscure Moravian monk fond of peas. By cross-pollinating pea plants, he shows that traits pass from both parents to their offspring yet remain fully separate. They can be dominant, recessive, homozygote, or heterozygote. What ridiculous complexity!

The homunculus idea was nice and simple, but couldn't explain why offspring often look like their mother. The blending of traits sounded great, too, but would inevitably kill off variation, because the entire population would converge on some average. At first, the monk's work was criticized, then ignored and forgotten. Science was simply not ready for it. Fortunately, it was rediscovered decades later. The scientific community compared ideas, looked at evidence, listened to arguments, and began to favor the monk's explanation. Since his experiments were successfully replicated, Gregor Mendel is now celebrated as the founder of genetics.

In comparison, religion is static. It does change with a changing society, but rarely as a result of evidence. This sets up a potential conflict with science, such as the never-ending clash about evolution. The real point of contention in this particular case is relatively minor, however, at least for the biologist. How humans relate to the rest of nature is hardly the core of evolutionary theory, yet it constitutes the main stumbling block for religious detractors. One rarely hears objections to the evolution of plants, bacteria, insects, or other animals: it's all about our own precious species. If we weren't put on earth by God, so the thinking goes, we'd lack purpose. To understand this obsession with human origins, keep in mind that the Judeo-Christian tradition arose with little or no awareness of other primates. Desert nomads knew only antelopes, snakes, camels, goats, and the like. No wonder that they saw a yawning gap between human and animal, and reserved the soul just for us. Their descendants were shocked to the core of their beliefs when, in 1835, the first live anthropoid apes went on display at the London Zoo. People were offended, unable to hide their disgust. Queen Victoria judged the apes "painfully and disagreeably human."[18]

Human exceptionalism is still very much alive in the social sciences and the humanities. They remain so resistant to comparisons of humans with other animals that even the word "other" bothers them. The natural sciences, in contrast, having suffered less religious contamination, march inexorably toward ever greater human-animal continuity. Carl Linnaeus placed *Homo sapiens* firmly within the primate order, molecular biology revealed human and ape DNA to be nearly identical, and neuroscience has yet to find a single area in the human brain without an equivalent in the monkey's. It is this continuity that is controversial. If we biologists could just debate evolution without ever mentioning humans, no one would lose a night's sleep over it. It would be like our discussing how chlorophyll works or whether the platypus counts as a mammal. Who cares?

I was largely unaware of evolution skepticism before coming to the United States, and quickly categorized it with other incomprehensible national penchants, such as love of guns and contempt for soccer. That evolution denial is as American as apple pie became clear again in 2011 when forty-nine of the fifty-one Miss USA Pageant contestants answered the question "Should evolution be taught in schools?" by hedging their bets. The jury is still out on evolution, they said, and there are so many religious views and scientific theories that it's better to teach all sides. Miss Alabama even felt that evolution should flat-out not be taught at all. In a case of divine justice, the title was won by Miss California, who explicitly endorsed evolution, saying she was a "huge science geek."

No less than 30 percent of Americans read the Bible as the actual word of God. But this is still only half of those who feel that the Bible is either an inspired text not intended to be taken literally or a book of legends and moral precepts.[19] This is great to know for those trying to get an evolutionary message across. The nonliteralist majority is (or should be) their target audience, since they are most likely to listen. Except, of course, if the discussion opener is a slap in the face. Unfortunately, all this talk about how science and religion are irreconcilable is not free of consequences. It tells religious people that, however open-minded and undogmatic they may be, worthy of science they are not. They will first need to jettison all beliefs held dear. I find the neo-atheist insistence on purity curiously religious. All that is lacking is some sort of baptism ceremony at which believers publicly repent before they join the "rational elite" of nonbelievers. Ironically, the last one to qualify would have been an Augustinian friar growing peas in a monastery garden.

The greatest public defender of evolution this country has ever known was Stephen Jay Gould. During his heyday, Gould was so popular that he single-handedly carried an entire magazine, *Natural History*, which since his death has become only a shadow of its former self. Gould was always a pleasure to read, especially for his eye-opening excursions into science history, which he seemed to know like his back pocket. One doesn't need to agree with all of his opinions and every fact he touted—I most certainly don't—to recognize that he was the face of evolution and its foremost advocate. He wrote with such contagious enthusiasm that he inspired thousands of young Americans to go into science.

His defense of evolution included frequent warnings against the racism and genetic determinism associated with it in the old days. He also vehemently resisted the idea that every single human behavior deserves an evolutionary account. "Darwinian fundamentalism," as he disparagingly called it, holds that everything we do is controlled by genes and serves to propagate them. Gould explained that Darwin himself doubted such a sweeping role for natural selection, and I have already mentioned the many kinds of altruism that are left unexplained. Dismissing such behavior as "mistakes" doesn't solve much. And this isn't the only behavior that makes little evolutionary sense. Think of smoking, masturbation, bungee jumping, boozing, setting off fireworks, and rock climbing. Maladaptive behavior is in fact so common in our species that we make fun of it. The Darwin Awards were invented to honor "those who improve the species by accidentally removing themselves from it."

This is not to say that we shouldn't try to put human behavior on an evolutionary footing. There really is no alternative. Indeed, I predict that fifty years from now Darwin's portrait will hang in every psychology department. Yet, the field still teems with "just so" stories that are hard to take seriously, from the proposal that male-pattern baldness serves as a signal of wisdom to the opposite sex (something we need to tell all those men with an illusory head of hair), to *The Natural History of Rape*, which is the actual title of a book that has harmed evolutionary psychology more than any other. The big mistake, of course, is to assume that if something is genetic, such as baldness, it must be good for you. Alzheimer's, cystic fibrosis, and breast cancer all have a genetic basis, but no one would want to argue that they increase fitness.

In the case of rape, however, we don't even have a genetic basis to work with. There is no evidence whatsoever that sexual violence is heritable. Still, this didn't stop Randy Thornhill and Craig Palmer from speculating about its evolutionary benefits. Extrapolating straight from the sexual behavior of flies, the authors proposed that men rape in order to spread their genes. Worse, the authors absolved themselves from producing actual data, since most important effects must have occurred during human prehistory. With this past being a closed book, all we were left with was unrestrained storytelling. The authors never answered the question why, if rape is all about procreation, one-third of its victims are nonreproductive, such as children and the elderly.[20]

I agree with Gould that we gain little from evolutionary guesswork about each and every human behavior. Gould made himself many enemies, though, by voicing skepticism. Several skirmishes between him and the evolutionary establishment unfolded in the pages of the *New York Review of Books* in 1997. It was a sight to behold, all those outsized egos tumbling over each other with innuendos, criticism by hearsay, name-calling (one was ridiculed as another's "lapdog"), or acting as if they'd never heard of one another. The vitriol obviously didn't help them close ranks. Creationists were rubbing their hands in delight, and exploited the row to their own ends as was reflected in remarks about theoretical discord in places as unexpected as the Miss USA Pageant. Someone should have nominated those Darwinists for a Darwin Award.

But this confrontation was nothing compared with the reaction to another Gould opinion. An atheist himself, he declared science and religion compatible well before neo-atheists decided they were not. Following his untimely death, in 2002, Gould thus became a lightning rod for his lack of intolerance.

Somethingism

In his famous essay, published in the same year as the above lovefest with fellow evolutionists, Gould recalled running into a group of lunching priests at the Vatican. The priests expressed worry about a new brand of creationism that had sprung up, known as *intelligent design*. They asked Gould why on earth evolution was still under attack. In his essay, the paleontologist commented on the profound irony that he, an ex-Jew, had to reassure Catholic priests that evolution was in fact doing fine and that the opposition was restricted to a small segment of the American population.

This story was Gould's way of hinting that the alleged war between science and religion is overblown. Blanket statements about "religion" are problematic anyway, since the term covers everything from monotheism to polytheism, and from a rigid belief system to spirituality. Buddhism, for example, actually welcomes the idea of evolving organisms, which agrees perfectly with its view that all life is interconnected and in flux.[21] But even within religions, such as Christianity or Islam, cultural diversity is so great that practices and ideas abhorred in one corner are often supported in another. Indonesian Sunnis have about as much in common with Iranian Shi'a as Swedish Lutherans with Baptists in the American South. With a better grasp of these issues than most, Gould borrowed the term "magisterium" (teaching authority) from a papal document to make the point that science and religion occupy separate spheres of knowledge. They don't touch on the same problems. Speaking of "nonoverlapping magisteria," he designated this view NOMA.

We face two distinct sets of questions, one related to physical reality and the other to human existence. Given how little science tells us about the second question, a French biologist, Matthieu Ricard, turned his back on a promising career in science to become a Buddhist monk. I have met Matthieu several times, and it isn't hard to detect an inner peace in him that few people possess. On the basis of fMRI scans of his brain during meditation, he has been dubbed the "happiest man in the world" (a title he gives a Gallic shrug). Neuroscientists measured the highest activation ever in his left prefrontal cortex, which is associated with positive emotions. Even though Matthieu still talks with the precision of a scientist, he abandoned science long ago, arguing that all it delivered was "a major contribution to minor needs."[22] This echoes Leo Tolstoy's complaint that whenever he asked scientists about the meaning of life, such as what we should do with it, he "received an innumerable quantity of exact replies concerning matters about which I had not asked."[23]

Few scientists follow in Matthieu's footsteps, however. Instead of turning to religion, the majority of us are agnostic or atheist. This shouldn't be taken to mean that science answers questions of meaning and purpose, however. Even the scientists who recently confirmed the "God particle" knew that it was a far cry from confirming why we are on earth and even less whether or not God exists. No, the big difference for scientists is that the thirst for knowledge itself, the lifeblood of our profession, fills a spiritual void taken up by religion in most other people. Like treasure hunters for whom the hunt is about as important as the treasure itself, we feel great purpose in trying to pierce the veil of ignorance. We feel united in this effort, being part of a worldwide network. This means that we also enjoy this other aspect of religion: a community of like-minded people. At a recent workshop, a retired astronomer teared up while discussing humanity's place in the cosmos. He stopped talking for two minutes, causing his audience to become restless, before explaining that he had pursued this question since childhood. The sight of images from billions of light years ago still overwhelms him, making him realize how much we are connected with the universe. He wouldn't call it a religious experience, but it sounded very much like it.

Along with people in other creative professions, such as artists and musicians, many scientists experience this transcendence. I do so every day. For one, it's impossible to look an ape in the eye and not see oneself. There are other animals with frontally oriented eyes, but none that give you the shock of recognition of the ape's. Looking back at you is not so much an animal but a personality as solid and willful as yourself. This is a familiar theme among ape experts, who will tell you how their very first eye contact radically changed not only how they viewed their subjects but also their own place in the world. It is precisely this impact that upset Queen Victoria. Staring into the ape's mirror, she felt the metaphysical ground shifting underneath her royal feet. Seeing the same orangutan and chimpanzee at the same zoo, Darwin reached quite a different conclusion; he invited anyone convinced of man's superiority to come take a look. Darwin felt a connection where the queen felt a threat.

Watching a magnificent landscape or a sunset over the ocean, most of us feel like a minuscule part of the universe, the way scientists feel while looking through a microscope or telescope, analyzing whale song, digging up dinosaur bones, or running after chimps in the forest. The chimps dive into the undergrowth while their bipedal cousins struggle with a large bush knife to keep up with every turn they take, listen to every hoot, and document every social encounter. Accompanying my late friend Toshisada Nishida, a Japanese primatologist known for his fieldwork in Tanzania, I was struck by his habit of chewing every leaf or fruit he saw the wild chimps eat. He wanted to know how they tasted, he said, but for me it was the ultimate act of feeling one with a kindred species. The same identification was at play when a young British primatologist, Fiona Stewart, did what no one had done before: sleep in tree nests built by chimpanzees. Nests had always been studied from the ground, with binoculars, but Stewart spent the night in them. She discovered advantages over sleeping on the ground, such as a deeper sleep and fewer insect bites. Other scientists trail dolphins in speedboats, having given each one a name and recognizing them by their fins. Or they fly ultralight aircraft followed by fledgling whooping cranes in order to introduce them to the air. All of this rests on attraction to the natural world, which often started early in life. Specialized knowledge about a tiny section ties us to its grandeur and complexity, which spreads in all directions, across all degrees of magnification, and across endless time. We are in awe of the mysteries we seek to decipher, which deepen with every layer we peel away.

I therefore fully understand how a prominent cell biologist, Ursula Goodenough, could write a book entitled *The Sacred Depths of Nature*, or how Einstein could believe in Spinoza's God. Baruch Spinoza, the seventeenth-century philosopher from Amsterdam, picked up skeptical strands of thought that had been around in the Netherlands since the days of Bosch and Erasmus, crystallizing them into an impersonal God. Not the traditional omniscient father figure in the sky, but an abstract supernatural force tied to nature. He thus laid the groundwork for a rationalist worldview in which scripture represents not the word of God but merely the opinion of human mortals. His message was not well received, to say the least, and Spinoza was excommunicated from his Sephardic Jewish community.

Einstein subscribed to Spinoza's God, but had no hostility toward religion, saying about its pervasiveness that a belief seemed to him "preferable to the lack of any transcendental outlook of life."[24] As it was with Gould, tolerance is key. Dogmatism closes the mind, whether it is the blindness of biblical literalists for science or the self-righteousness of some atheists. A recent illustration is the palace coup against Paul Kurtz, writer of the internationally celebrated *Humanist Manifesto* and founder of the Center of Inquiry. The legendary eighty-five-year-old became persona non grata in his own organization for not supporting Blasphemy Day and other silly ways to mock religion. This is how Kurtz himself explained the situation:

They wanted to be hard on religion. Now, I don't like God. I think she's a myth. I don't think there's evidence for her. But, on the other hand, many people believe in religion. Although I believe in criticizing them, I don't hate them. I'm not nasty towards them. So there's a difference in how you deal with religion. Many of my colleagues, I call them former altar boys, so hated religion that they couldn't help expressing that.[25]

Kurtz's altar boy reference hints at the serial dogmatism mentioned earlier, which simply redraws the boundaries of intolerance. Anti-something movements will go the way of the dodo, however, unless they manage to replace what they dislike with something better. They will need to come up with a viable alternative. No secular movement can get around Tolstoy's questions. The increasingly secular Dutch even coined a word for this: "ietsism." "Ism" is the same as in English, and "iets" means "something." The typical "ietsist" doesn't believe in a personalized God, and follows no traditional religion, yet thinks there must be more between heaven and earth than what meets the eye. There must be something.

The enemy of science is not religion. Religion comes in endless shapes and forms, and there are tons of faithful people with an open mind, who pick and choose only certain parts of their religion and have no issue with science whatsoever. The true enemy is the substitution of thought, reflection, and curiosity with dogma. The God debate in Puebla was intellectually dishonest and sanctimonious on both sides. Where do people get convictions stronger than anything I have ever experienced in my life? What is their secret? Convictions never follow straight from evidence or logic. Convictions reach us through the prism of human interpretation. As a French philosopher aptly summarized, "strictly speaking, there is no certainty; there are only people who are certain."[26]

Let me therefore close this chapter with a quotation from John Steinbeck, the American novelist, since it illustrates the other side: the seeking, wondering, and pondering human mind open to any and all influences. This is the mind that has driven us forward through the ages despite the intellectual ossification humans are prone to. Steinbeck depicts science and religion as equivalent forms of holistic knowledge. He probably would have agreed with Goodenough that the membrane separating Gould's two magisteria is "semi-permeable."[27] Like so many of our bodies' membranes, it lets chemicals through in both directions. After all, science has the potential to affect our social and moral outlook, such as when it promotes environmental awareness or invents a pill that offers women sexual freedom. Conversely, existential questions feed into science, as in the debate about humane versus medical considerations in the treatment of patients. "Should we keep everyone alive for as long as we can?" is not a question that science can answer. In many areas, it is hard to tell where our worldview ends and science begins, and vice versa. We need to step beyond a simple dichotomy between the two and consider the whole of human knowledge. Steinbeck tried his hand at it in the following passage from *The Log from the Sea of Cortez*, about a scientific expedition along the Pacific Coast:

> *And it is a strange thing that most of the feeling we call religious, most of the mystical outcrying which is one of the most prized and used and desired reactions of our species, is really the understanding and the attempt to say that man is related to the whole thing, related inextricably to all reality, known and unknowable. This is a simple thing to say, but the profound feeling of it made a Jesus, a St. Augustine, a St. Francis, a Roger Bacon, a Charles Darwin, and an Einstein. Each of them in his own tempo and with his own voice discovered and reaffirmed with astonishment the knowledge that*

all things are one thing and that one thing is all things—plankton, a shimmering phosphorescence on the sea and the spinning planets and an expanding universe, all bound together by the elastic string of time. It is advisable to look from the tide pool to the stars and then back to the tide pool again.[28]

Chapter 5

THE PARABLE OF THE GOOD SIMIAN

The very sight of another's pain materially pains me, and I often usurp the sensations of another person. A perpetual cough in another tickles my lungs and throat.

—Montaigne[1]

Elephants are easily underestimated. I surely was guilty of doing so in regard to their use of tools. All I'd ever seen them do is pick up a stick to scratch their behind. I had also watched them throw dirt. This happened every time when jackdaws tried out their spring song on the fence around the elephant enclosure at a zoo where I worked. These birds are like the raven in La Fontaine's fable. Jackdaws may think they can sing—after all, the crow family belongs to the songbirds—but musicality is a matter of taste. The elephants would throw trunkfuls of dirt at the noisemakers to get rid of them.

I thought this was the maximum that elephants were capable of, because experiments had never shown much more. Scientists had offered pachyderms a long stick while placing food outside their reach to test whether they'd use the stick to retrieve it. This works well with primates, but elephants leave the stick alone. The conclusion was that they didn't understand the problem. It occurred to no one that perhaps we, the investigators, didn't understand the elephant.

Unlike the primate hand, the elephant's grasping organ is at the same time its smelling organ. Elephants use their trunk not only to reach food but also to sniff at it and touch it—their trunk, especially the sensitive tip, is full of nerve endings. With their unparalleled sense of smell, the animals know exactly what they're going for. Vision is secondary. As soon as an elephant picks up a stick, however, its nasal passages are blocked. Even if the stick gets close to the food, it still impedes feeling or smelling. It would be like asking us to reach for something blindfolded. Except in a party game, we are reluctant to do so, and for good reason.

On a recent visit to the National Zoo, in Washington, Preston Foerder and Diana Reiss showed me what Kandula, a young elephant bull, is capable of if the problem is presented differently. The scientists hung branches with fruits high up above the enclosure, just out of reach of Kandula. The elephant was given several objects to work with, including sticks, a square box, and several thick wooden cutting boards. Kandula ignored the sticks, but after a while started kicking the box with his foot. He kicked it many times in a straight line until it was right underneath the branch so that he could stand on top of it with his front legs, which allowed him to reach the food with his trunk.

While Kandula munched his rewards, Preston and Diana explained that they had begun making his life more difficult by displacing the objects. They would put the box in a different section of the yard, out of view, so that when Kandula looked up at the tempting food he'd need to recall the solution and walk away from his goal to fetch the tool. Not many animals are capable of this, but Kandula did so with-

In order to reach green branches hung high above his head, Kandula had to find and fetch a box to stand on, which he did.

out hesitation, retrieving the box from great distances. And when the box was removed entirely, he picked up the wooden boards to stack them on top of one another in order to get closer to the food.

Kandula showed all the signs of cause-effect understanding, also known as the "eureka! moment," that is considered a sign of high intelligence. Clearly, we should first try to look at the world from another animal's perspective—even if this means picturing ourselves outfitted with a hose as a nose—before claiming that throwing dirt is all they are capable of.

Another's Welfare

This recalls an earlier time when Diana and I collaborated to see whether elephants recognize themselves in the mirror. Together with Josh Plotnik, then a student of mine, we conducted a study at New York's Bronx Zoo. Elephants had never before shown any signs that they knew what they saw in a mirror. Did they think it was another elephant, the way monkeys see another monkey? Only humans, apes, and dolphins were known to recognize their own reflection.

Previous tests, however, had presented the largest land animal with a mirror far smaller than itself. This mirror had been placed at ground level outside the bars of an indoor cage. All that the elephant might have seen was four legs behind two layers of bars, since the mirror doubled them. The disappointing outcome had been taken to mean that elephants don't recognize themselves. We wondered, however, if there wasn't a better way of testing them. An expensive eight- by eight-foot elephant-proof mirror was put inside their outdoor enclosure, so that they could feel, smell, and look behind it before exploring their reflection. Exploration is a necessary first step, also with chimps and children. Lo and behold, one Asian elephant, named Happy, recognized herself. She repeatedly rubbed a white cross on her forehead while standing in front of the mirror. She could know about this cross only by connecting her reflection with her own body. This is also the way self-recognition is tested in children, who usually show it before the age of two. For elephants to join the self-aware animal elite was a big deal. The media headlines couldn't resist reference to a children's song: "She's Happy, and she knows it!"

Elephants turned out smarter than had been thought, but most importantly the finding confirmed the limits of negative evidence. If you don't find tool use or signs of self-recognition in a given species, you still don't know what's going on. It could be that the animal is not up to the task, but it could equally well be that we are not up to the animal. We may be giving it the wrong tools or holding up the wrong mirror. This insight is reflected in the famous dictum of experimental psychology that "absence of evidence is not evidence of absence."

It is a point that bears repeating. If I walk through a forest here in Georgia and fail to see or hear the pileated woodpecker, am I permitted to conclude that this bird is absent? Of course not. I may just have missed it. We know how easily these splendid woodpeckers hop around tree trunks to stay out of sight. As big as crows, they glide through a woodlot like a vision, which is why early woodsmen called

As a three-year-old male bonobo at the Cincinnati Zoo (above), Vic befriended a same-aged boy, who was brought by his father every week to see him. The ape and the child would stare through the glass into each other's eyes, and were clearly friends, according to the photographer Marian Brickner. She says, "What a dad! I never got his name. I got the picture, actually had to ask him to please step to the left so I could get the right angle." In 2012, Marian returned to the zoo to obtain an updated portrait of Vic, who is now twelve years old, which makes him a late adolescent (right).

Bonobos have sex in all positions and all partner combinations. Face-to-face copulations are common, such as here by an adult opposite-sex pair (above). Same-sex couplings are common as well, most typically between females. One adult female invites another for genital rubbing (right). Note the role of eye contact, which is actively sought and maintained during sex.

Mama with her daughter Moniek. For me, this female remains the quintessential monarch of chimpanzee society as she ruled the large Arnhem Zoo colony for decades, even when she could barely walk anymore. She did not physically dominate any adult males, but everyone reckoned with her influence, and males would literally seek refuge in her arms if their rivals were putting too much pressure on them.

The resemblance between ape facial expressions and human ones is illustrated by this laughing chimpanzee. While holding my camera aloft with one hand, I tickle this three-year-old male in his side with the other. He utters hoarse guttural sounds with the same rhythm as human laughter but without the high-pitched shrieks that our species produces. There is no mistaking the sense of fun as he begs for more.

Community concern is reflected in attempts to restore harmony. A high-ranking male engages in impartial arbitration after two females clashed over browse food (top, center). With outstretched arms, he stands between them until the screaming stops. Females also intervene, such as Mama (bottom, center) in a conflict between the alpha male and a juvenile. Mama approaches the male with appeasing pant grunts, then guides the juvenile away from the scene.

A code of conduct is dictated by the need to get along. Mutual trust is implied during play when young chimps become fully entangled, gnawing on each other's hands and feet (top). If the game gets too rough, amends will need to be made in order for the fun to continue. Violations of social rules are protested—for example, by a screaming young chimp begging with outstretched hand for the berries someone stole from him (bottom).

A rich repertoire of reassurance gestures serves to calm others and oneself at times of anxiety. Two female chimps embrace while watching a tense fight within their community (top), and a female makes up with a dominant male, after a fight between them, by kissing him on the mouth (bottom).

Adopting the young of others is a common form of altruism in both humans and other animals. A female chimp with deficient lactation had lost several offspring before I taught her how to bottle-feed an adopted baby, "Roosje" (right). She raised not only Roosje on the bottle but her own subsequent offspring as well. My lessons earned me this female's lifelong gratitude.

them the "gawd-almighty." These birds are shy most of the time, however, and their laughing calls and drumming are heard only in certain seasons. If after multiple strolls, I still haven't noticed them, all I can say is that I lack evidence. Perhaps the gawd-almighty isn't around, but I wouldn't bet my life on it.

It's quite puzzling, therefore, why the field of primate cognition has such a long history of claims about absent capacities that are based on just a few strolls through the forest. The most recent one was that apes aren't prosocial. They resemble us in many ways, but, poor things, they are as selfish as a pickpocket. All they care about is themselves. The Prosocial Choice Test, which I will describe in a moment, declared chimpanzees to be insensitive to the interests of others. No matter that they show plenty of spontaneous helping, such as food sharing or risky defense of friends, bite through poacher snares when group mates are trapped, and adopt unrelated orphans. No matter, moreover, that they help others retrieve a needed tool or open a door so that a companion can reach food.

All very nice, the academic community said, but so long as apes flunk the critical test of generosity, we don't believe a word of it. Hence the oft-repeated claim that chimpanzees base social choices "solely on personal gain,"[2] that human cooperation represents a "huge anomaly" in the animal kingdom,[3] and that prosocial tendencies evolved as a "derived property of the human species" after our ancestors split off from the apes.[4]

This negative view of the apes held for about a decade, until my team tested the chimpanzees that I am watching at this very moment. I'm sitting behind my desk in an office that overlooks the field station of the Yerkes National Primate Research Center, outside of Atlanta. The chimps literally live below my window. I have occupied this office for over two decades, hence every chimp in its twenties was known to me as a youngster. This includes the current alpha male, Socko, and alpha female, Georgia, both of whom I tickled when they were little. They used to laugh their hoarse laughs while begging for more. The group used to include Peony, the previous alpha female, who by the end of her life was assisted by others, who brought her drinking water, helped her climb, and so on. No wonder that these chimpanzees consider me and my tower office part of their territory. If I bring visitors, everything is fine, but people who come here on their own aren't always welcome. One time, I arrived after a rainy period, which had turned the enclosure into a mud bath, and found my whole window covered with chunks of dried dirt. I didn't understand it, until someone explained that a cleaning crew had spent time in my office to the displeasure of the chimps.

In the same way that no one would judge the mental capacities of children on the basis of what they do in the play yard, chimpanzee intelligence cannot be understood from mere observation. Fieldwork has its place and value, but the assessment of ape cognition requires that we present them with specific problems. We conduct many such tests, following a two-point philosophy. First, we focus on what chimps can do, not what they can't do. Given the problems with negative evidence, we just don't dwell on it. Second, we keep our experiments simple and intuitive. This is where experience with spontaneous behavior comes in handy, such as what I see them do every day looking up from my desk. We design tasks that chimps are interested in and find easy to handle. For example, instead of testing their willingness to imitate humans, we'd rather see what they do with each other. Chimps are not nearly as interested in us as we would like them to be, yet they eagerly pay attention to each other. They press their face close to that of another to follow each and every move, smell another's mouth, and even lay their hand on that of another who is performing a task, thus gaining kinesthetic feedback. Apes aren't particularly good at imitating humans, but we found that they love to ape apes, which is all that matters.

Our chimp-friendly approach served us well while developing a new Prosocial Choice Test. The test that chimpanzees had repeatedly failed involved an apparatus that delivered food to either one or both members of a pair. An ape was given the choice between pulling a lever to obtain food for himself or a different lever to reward both himself and a partner. This means that he could advantage the other without a cost to himself. The chimps pulled both levers equally, however, as if they didn't care.

But, obviously, it all depends on how well they understood the apparatus. Did they know how their choices affected their partner? We were skeptical. The way we often go about designing studies is to sit around during morning coffee. I lead a small team of a dozen graduate students, postdoctoral scientists, and technicians. One member of the team will explain his or her experiment, and all of us will jump in with criticism, explaining why something might not work, suggesting alternatives, citing other studies, and so on. We go around many times, sometimes for months, until we have a design that seems at the right level for the apes. The same happened with the prosocial project. We first went over the details of previous studies. We noticed, for example, that the apes were often placed meters apart with more than one layer of glass between them. Producing food for themselves must have been clear enough, but could they tell what their partner was getting? We looked at pictures of the apparatus and were astonished by its complexity. If we couldn't figure out how it worked, wouldn't chimps have trouble, too? We set out to eliminate all of those issues.

Vicky Horner, a scientist who knows chimpanzees inside out from both the field and captivity, modified a design used earlier with monkeys. It avoided an apparatus altogether. We placed the chimps close together, with just one layer of mesh between them, and wrapped their rewards of banana slices in butcher paper. This made it about as hard to eat in silence as it is to unwrap a bonbon unnoticed during a classical concert. We wanted to be sure the chimps knew about each other's food consumption. I remember the day we settled on the final plan, which we tried out in a mock-up with us as players. We imagined being chimps and going through the motions. Since we felt comfortable with the procedure, it was time to get to the proof of the pudding.

Our chimps live in a spacious enclosure with grass and wooden climbing frames. We call them by name to enter a building, where we separate them according to plan. Everything is done on a volunteer basis, but since food is involved, they are eager. We face problems, however, that human psychologists never deal with. For example, if a female's genitals are swollen, she is a hot commodity for the males, who either refuse to let her go, by blocking the entrance to the building, or stay outside while she enters but bang on the metal doors during the entire test. The amount of noise a motivated male can produce is frightening, and obviously doesn't help concentration. Or a youngster may resist separation from its mother, coming in with her, so we can't test her alone. Then there are all the colds and illnesses, fights, bad-weather days, and other interferences. Since we need to test multiple chimps multiple times before our results are conclusive, a study like this can easily take a year.

The effort paid off, however, because we were the first to demonstrate that chimps care about each other's welfare. Vicky's persistence, the new test design, and our rapport with the apes all came together. The very first test involved Peony and Rita, two unrelated females. Once they had settled down in separate rooms, we filled a bucket with colored tokens. These were small pieces of plastic pipe, all identical, except that half of them were green and the other half red. We asked Peony to pick one token at a time and hand it to us. Whether she picked green or red, she was always rewarded. The only difference concerned what Rita received. Red tokens were "selfish" in that they rewarded only Peony, whereas green tokens were "prosocial" in that they rewarded both Peony and Rita. Choosing many times in a row, Peony began to select green tokens two out of three times. In a few other chimpanzee pairs, we observed nine out of ten prosocial choices,

In front of a waiting partner, the acting chimp (right) reaches into a bucket with plastic tokens of two colors. After the actor has selected a token, we place it on a table between two paper-wrapped rewards. We reward either only the actor ("selfish" color) or both apes ("prosocial" color). Chimpanzees develop a preference for the prosocial option.

but on average the inclination to help the other was in line with Peony's: not very high, but not random either. If we placed a chimpanzee alone in the room, on the other hand, and applied the same procedure, the two colors were treated without distinction. In other words, the prosocial preference required a partner.[5]

The partner obviously knew what was going on and tried to influence the chooser by intimidating her, banging on the mesh between them, spitting water, hooting loudly, or begging with open hand. Partners did so mostly after selfish choices. Such pressure was counterproductive, however, in that it caused a drop in prosocial choices. It was almost as if the choosers were telling their partners to behave; otherwise they'd get nothing. Having tested twenty-one different pairs of chimpanzees, we ruled out fear as an explanation, because the highest-ranking chimps, who had least to fear, were the most prosocial. Both Peony and Georgia acted generously.

Peony's bigheartedness didn't surprise us in the least. All of her life, she had been the sweetest character, ready to help and reassure everyone, which also explains the love and respect she received in return toward the end of her life. Georgia was a different story. She is known as a bully and troublemaker. When she turned sexually mature, she'd instigate fights among males by having sex with low-ranking ones in full view, or she'd hit the children of other females, thus provoking brawls that could involve half the community. I have never looked at Georgia as a generous character, therefore, but in our tests she proved to be. Was she, unbeknownst to us, good at doling out favors? Had Georgia's rise been built as much on the carrot as the stick? Among females, such strategies are harder to detect than among males, who make a big show of every intimidation or favor.

Most chimpanzee experts reacted to our study with relief. They had watched in disbelief how negative data seemed to be winning the day. Some had gone so far as to collect fecal and hair samples in the wild in order to extract DNA to demonstrate that chimpanzee cooperation isn't just a matter of blood relations. It's quite common for nonrelatives to help each other. But apart from pleasing the experts, our findings also faced some grumbling. One economist complained that since the chimpanzees faced no cost to their behavior, it didn't really count as altruism. Another pointed out that the chimps weren't prosocial *all* the time. Since they ignored their partner's interest one out of three times, they weren't particularly nice, he said, calling them "mean-spirited" instead.[6]

Now, remember that for an entire decade humans had been considered unique because chimps, it was said, didn't even help each other when doing so didn't cost them anything. We improved the methodology, got rid of the apparatus, and, voilà, our chimpanzees did help each other. For colleagues now to complain that this didn't really count is a bit rich. Clearly, we need more research, but previous claims are now in doubt. Children subjected to the same Prosocial Choice Test are imperfectly prosocial, too: in one study, they helped each other 78 percent of the time. The difference with chimpanzees is therefore one of degree.

Goalposts move magically as soon as humanity's special status is at stake. Prosociality in other animals has become hard to deny, however. Apart from the studies mentioned, a new factor is the ubiquity of video cameras. It used to be that we had to rely on accounts from the field, such as Goodall's story of how Madame Bee had become too old and weak to climb into fruiting trees. The aging chimp would patiently wait for her daughter to come down carrying fruits, one of which she'd place beside Madame Bee, upon which the two of them would contentedly eat together. We believed this kind of report without further evidence, but now we have dozens of intriguing videos on the Internet. Millions have watched the "Battle at Kruger" on YouTube, in which a buffalo herd saves a calf from lion claws, or the "Hero Dog," who risks his life dragging a half-dead comrade off a busy highway in Santiago, Chile. There is also the dog who refuses to leave an injured companion after the recent Japanese tsunami, African elephants who pull a calf from a mud hole, and beluga whales at a Chinese zoo who save a human diver who stopped breathing at the bottom of a freezing tank. One beluga gently takes her in its mouth, the way a dog picks up a puppy, after which both whales push the diver to the surface.

With contrary evidence so vividly on everyone's computer screen, the idea that only humans care about the welfare of others is rapidly losing steam.

Georgia's Gratitude

You can't trust everything you see on YouTube, though. Another video shows synchronized swimming by goldfish. Four fish move together like a military jet formation, keeping exact distances from each other, turning corners at the same time while moving across a shallow tank. They seem orchestrated by the hands of a magician floating above them. The video caused a public outcry when it was speculated that the fish had been fed small iron pellets. They may have been directed by someone with magnets underneath the tank, which would explain its shallowness. Without denying the accusations, the magician claimed that his fish were happy and healthy.

The whole performance was unlikely to begin with, because carp aren't known for tight schools. All they do is aggregate. Animals that normally don't do much together are almost impossible to train on coordination. One can teach two dolphins to jump simultaneously out of the water precisely because dolphins do such things spontaneously. Female dolphins swim synchronously with their calves, surfacing and diving together. Male dolphins form coalitions that intimidate rivals by swimming in perfect synchrony, thus showcasing the tightness of their bond. Try to get two domestic cats to jump through a hoop together, however, and you are in for failure. Cats are solitary hunters.

In our own studies of cooperation, we avoid any specific training. We want to know how well animals understand the concept. Do they have shared goals? Do they recognize the partner's effort? One of my students, Malini Suchak, is setting up an experiment that requires chimpanzees to work together. Instead of testing them in pairs, as is tradition, we conduct the project outdoors in the presence of the whole colony. This forces them to recruit each other, similar to wild chimps setting out on a monkey hunt. Moving targets are hard to catch in three-dimensional space, which is why chimps hunt more successfully in pairs or trios than alone. I have witnessed them chase monkeys high up in the canopy, culminating in a chorus of excited screams when they make a capture. Hunting and meat sharing are at the root of chimpanzee sociality in the same way that they are thought to have catalyzed human evolution. The big-game hunting of our ancestors required even tighter cooperation.

Malini's apparatus is mounted on the outdoor fence. By now, the chimps know that on their own they can't get any goodies from it. Sitting at it, Rita looks up at her mother, Borie, who is asleep in a nest she built on top of a tall climbing frame. Rita climbs up all the way to poke Borie in her side until she comes down with her. Rita heads to the apparatus, all the time looking over her shoulder to make sure her mother is following. At other times, we have the impression that the chimps have reached an agreement without our noticing. Two of them will walk side by side out of the night building and head straight for the apparatus, as if they know exactly what they are going to do.

Chimpanzees are masters at communication through subtle glances and body postures. Without language, and often without explicit gestures, they make clear what their next move will be. Their reliance on body language makes them excellent signal readers of humans. They are in fact so good that they seem to know my own moods and intentions better than I do myself. It's as if they look straight through us. I have also often noticed how sensitive they are to people with hostile body language, such as visitors who arrive with prejudices. In our presence, these people obviously don't mock the chimps the way zoo visitors sometimes do (hollering, exaggeratedly scratching themselves), but the chimps nevertheless react as if they've seen the enemy. Georgia will sneak off to collect a mouthful of water from the faucet and mingle with the rest, so that no one suspects the imminent waterworks. Human respect is also noticed, however. When I showed the veteran fieldworker Toshisada Nishida around, the apes did not react at all. He was standing next to me, leaning sideways a little, walking quietly without any abrupt moves as I had also seen him do in the forest, and the apes seemed to think that this man was perfectly all right.

Georgia was once removed from the group for eighteen months. Her removal was prompted by political instability. The males had been fighting, and the females, too, so much so that we feared for the lives of some juveniles. Low-ranking mothers had trouble prying their offspring from the clutches of adolescent kidnappers who enjoyed the backing of higher-ups. Ape babies nurse for four years, which is why involuntary separations from their mothers are worrisome. Several individuals were removed to calm down the group. Some permanently, but most were gradually reintroduced. Georgia was last. Despite her poor reputation, I kept insisting on her return. She was born into this group, she had good relations with the majority, and one day she might mellow and become a model citizen. I remember the looks of disbelief at my optimism, but after much back and forth, she was allowed to return.

With many animals this isn't a good idea. Rhesus macaques, for example, don't take well to a former troop member after a long absence. It is as if their position has been filled. These monkeys have a strict dominance hierarchy, so if the individual at rank 10 is removed and returned, even after just a few months, rank number 10 has been claimed in the meantime, and all other ranks as well. Hostility to the returning monkey is almost as fierce as that to a stranger. With chimps, however, there is no problem. Their hierarchy is a loose construct, and they live in so-called fission-fusion societies, in which parties meet, fuse, and separate all the time. Wild chimps may go months without seeing another member of their community.

Georgia's return was triumphant. Despite all of the excitement, and the occasional pushes and shoves she received, she stood her ground as if she had been gone for just a few days. She embraced and groomed with whoever was ready for it, and watched the males compete over who could sit next to her. Her mother and sister carried and protected Liza, Georgia's four-year-old daughter, and she herself reintegrated so quickly that by the end of the same week one could barely tell that she'd been away. She acted dominant to the females she had dominated before and was in every way her own self-assured self.

When I approached to take a close look at her and Liza, something happened that I remember to this day. When Georgia was young and playful, we had a fine relation, but as an adult she started to ignore me. We were largely on neutral terms. Now, however, she walked up to me and looked me straight into the eyes. Her gaze was positively friendly. She reached out to me. When I took her hand, she panted at me in a quick rhythm, which is about the kindest sounds a chimp can make. She did this only that time: never before, never since. It was not just a greeting, because I had visited Georgia multiple times in between without any such behavior. Since it happened right after her reintroduction, the two must have been connected. She may have noticed how glad I was to see her back. Perhaps it went even further, and she had detected tensions over her fate and my advocacy of her case. As I've said, chimps are incredibly astute readers of bodies and voices. We will never know, but I felt she was thanking me for getting her back to where she belonged.

I chalk this up to expressions of *gratitude* of which there are many. One concerned a different chimp whom I had taught to bottle-feed a baby. She had lost several offspring because of deficient lactation and was eager to adopt. Her previous losses had caused deep depressions, during which she'd isolate herself and scream for no apparent reason. Apart from raising this young chimp, she was able to keep her own infants in subsequent years thanks to her special skill. For a tool-using animal, giving the bottle is really not that hard. The rest of her life, this female was always insanely excited to see me, which happened only once every couple of years, as if I was a long-lost family member. It seemed related to my having helped her start a family. Yet another illustration of gratitude is the telling anecdote by Wolfgang Köhler, the German pioneer of ape tool use. Two chimps had been shut out of their shelter during a rainstorm when Köhler happened to come by and found the apes soaking wet, shivering in the rain. He opened the door for them. But instead of hurrying past him to enter the dry area, both chimps first hugged the professor in a frenzy of satisfaction.

Gratitude helps us render another person his due. Since it keeps favors flowing, it is essential for a society based on reciprocity. It was so highly regarded by Thomas Aquinas that he called it a secondary virtue, tied to the primary one of justice. Gratitude creates a warm feeling about received benefits, which prompts us to repay them. Why else would we do so? Out of duty? It is so much easier if our memory predisposes us kindly toward a benefactor. We'd barely feel it as a repayment. This is why Robert Trivers, the architect of reciprocal altruism theory, proposed gratitude as a critical ingredient.

Instead of relying on stories like those above, however, we have actually measured the exchange of favors. On many days throughout the year, I recorded grooming among our chimps in the morning before arranging a food-sharing session in the afternoon. I'd cut branches with leaves in the forest around the field station (chimps love young blackberry shoots and sweet gum) and tie them together with honeysuckle. Two large bundles of browse would be thrown into the compound. Any adult of any rank could claim and keep a bundle, because chimps respect ownership. Soon the lucky possessors would be surrounded by a circle of beggars holding out a hand, whimpering and whining. Even the highest-ranking male could be seen begging, as has also been reported from the field when chimps surround a monkey carcass. In the end, everyone would be munching on food obtained either directly from the owners or indirectly via family and friends. Having measured nearly seven thousand food interactions over many different sessions, my data showed a link between access to food and earlier grooming. On a day on which Socko had groomed May, for example, his chances of getting a few branches from her went up markedly compared with days on which he had not groomed her. This strongly suggests that chimpanzees remember and appreciate previous favors.

Reciprocity also works on the negative side, as predicted by Trivers, who saw a role for *moralistic aggression*. We are outraged at those who gladly take benefits without giving much back. Similarly, chimps may turn on their own allies if these decline to support them in a fight with another chimp. They stretch out a hand to a good friend on the sidelines, trying to get him to come over and stand shoulder to shoulder against an adversary. But the friend walks away. The jilted individual may interrupt the confrontation he was involved in to go after his so-called friend, screaming at the top of his lungs. All of this happens in noisy and tumultuous situations—there is nothing more chaotic and nerve-racking than a large chimpanzee fight—but such reactions help keep reciprocity on track. Chimpanzees also retaliate. If they have lost a fight to an alliance of several others, they may wait for the right occasion to get even. Encountering one of the others alone, without any of his buddies around, they will start a fight. I have known males who were very systematic about this: if they had been defeated by a quartet of opponents, they would take time in the ensuing days to have four unpleasant sessions with each one of them separately. More typically, however, a chimp defeated by an alliance waits until one of its tormentors is losing a fight against someone else, which is the perfect occasion to join the fray and add to the tormentor's defeat.[7]

Chimpanzee society always strikes me as revolving around tit for tat. These apes build a social economy of favors and disfavors ranging from food to sex and from grooming to support in fights. They seem to maintain balance sheets and develop expectations, perhaps even obligations, hence their negative reaction to broken trust. Because I am so used to how this works in our close relatives, I was struck by a nonreaction in another highly social animal. It happened while testing elephant cooperation in a Thai sanctuary with Josh Plotnik, who also conducted the earlier mirror study. We couldn't apply the usual technique, which is an apparatus so heavy that it forces individuals into joint action. With elephants, this would have required an apparatus the size of an eighteen-wheeler! Instead, we borrowed an ingenious design from a Japanese colleague, which consists of a single rope led all around an apparatus with both ends facing the animals. If they pull at just one end of the rope, it will unthread, come loose, and be useless. Only if both ends are picked up at the very same moment can the apparatus be pulled closer. Its weight hardly matters, but synchronization is a prerequisite.

It turned out a simple task for elephants. They would walk side by side to the two rope ends, pick them up, and pull. So far, so good. But then we began to introduce complications to see how well they understood the need for a partner. We would delay one of the two to see whether the other would have the sense to wait. The elephants showed impressive patience, waiting up to forty-five seconds. We also would remove the rope on one side, so that one of them couldn't pull. We found that in such cases the other didn't bother to try. He seemed to understand that pulling would be pointless.

Some elephants developed "illegal" practices, sidestepping our intentions. For example, one young cow would walk up to the rope and firmly put her big foot down on it while waiting for the other to arrive. This way, she didn't need to pull. Her foot would keep her end of the rope in place while her partner did all the work. She would not neglect, however, to reach into the food bucket once the job was done.

We took this as a sign not only of intelligence but also of cheating. Curiously, the other elephant never protested and never declined to pull. I'm not sure this would have worked so well with chimpanzees,

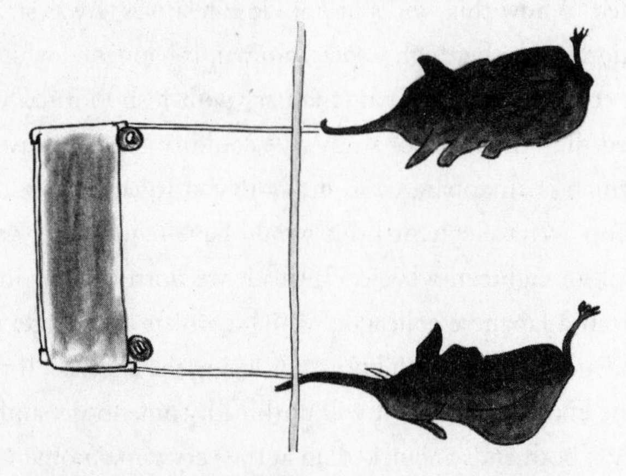

Viewed from above, two elephants pull at a rope to bring in a sliding tray with food buckets attached. They must pull at exactly the same time; otherwise the rope will unthread, leaving them empty-trunked.

which is exactly what Malini and I hope to find out in our cooperation study. Do apes and elephants differ because the latter have no idea of the contributions by others, or is it because the required effort is so minimal? The reward for each elephant was two cobs of corn—peanuts for such a large animal—and all they needed to do was pull a rope. Both effort and reward may have been insufficient for them to worry about freeloading.

All the better for us. Given the danger of working with elephants, the last thing we needed was irritated ones!

Body-to-Body Empathy

When Michael Jackson dangled his nine-month-old baby over the railing of a fourth-floor hotel balcony, holding the struggling boy awkwardly in just one arm high above the ground, many fans in the crowd below cheered, but some screamed. They were afraid Jackson might drop the baby, who—to make matters worse—had a towel over his head. It was the oddest scene, spoofed by Eminem and criticized by child abuse experts.

But why did we care? It wasn't our baby. We react this way owing to a capacity that was until recently largely overlooked by science: empathy. And here I don't mean empathy in the sense of sympathy, which urges us to assist others, but in the more neutral and fundamental sense of how we relate to others. When sixteenth-century French philosopher Montaigne said his throat itched as soon as he heard someone cough, he gave us the essence of empathy several centuries before the term came into use. Empathy connects bodies with bodies. Had we merely read about Jackson's baby handling in a newspaper, we might have shrugged it off as clumsiness, but we actually saw it on television. We saw how high-up the balcony was and noticed the father's tenuous grasp and the wriggling baby. Identification pulled us into the scene as if we, too, were holding the infant and felt its resistance. We lived the situation at a bodily level, which made it far more unsettling than any written account could have been.

Body mapping is automatic. The movie *The King's Speech*, for example, would be incredibly boring, were we to lack the ability to identify with others. Who cares whether a given word will be spoken quickly, slowly, or not at all? Only by feeling one with the king do we gain a stake in his speech problem, as if we suffered ourselves. Sitting at the edge of our seat, we'd like to say the words for him, willing him to succeed in the same way that parents make chewing movements while spoon-feeding their babies or mouth every word their child is supposed to say at a school play. We have this wonderful capacity to inhabit the bodies of others.

Putting it in neuroscience language, we activate neural representations of motor actions in our own brain similar to the ones we perceive or expect in the other. That we do so unconsciously has been tested with facial expressions on a computer screen. Even if the expressions are flashed too briefly for conscious perception (the subjects believe they're watching landscapes), our facial muscles still move along, and our mood is affected by the expressions seen. Frowns induce sadness, smiles happiness. Ulf Dimberg, the Swedish psychologist who conducted this research, told me about the initial resistance, which made it hard to get his findings published in the early 1990s. In retrospect, after so many confirmations, this seems absurd. But at the time, empathy was viewed as a complex skill under cerebral control. We *decide* to be empathic, so the thinking went, on the basis of deliberate simulations in our head of how we would feel in someone else's situation. Empathy was seen as a cognitive skill. Now we know that the process is both simpler and more automatic. It's not that we lack control (breathing is automatic, too, but we are still in command), but science looked at empathy entirely the wrong way. Empathy stems from unconscious bodily connections involving faces, voices, and emotions. Humans don't decide to be empathic; they just are.

True, we are capable of putting ourselves into someone else's shoes even in the absence of any bodily clues (such as when we read about a character in a novel), yet this doesn't make it the essence of empathy. To see the essence, consider how it starts. Look at a toddler who bursts out crying when her friend falls and cries, or who laughs heartily along with a room full of adults amused by a risqué joke far beyond the child's comprehension. Empathy finds its origin in bodily synchronization and the spreading of moods. Complex forms based on imagination and projection grow out of this, but only secondarily.

At about the same time as this groundbreaking research, scientists in Parma, Italy, discovered mirror neurons. These neurons are activated when we perform an action, such as reaching for a cup, but also when we see someone else reach for a cup. Since mirror neurons don't distinguish between our own behavior and that of others, they let one organism get under the skin of another. No wonder that their discovery has been hailed as being of equal importance to psychology as the discovery of DNA has been for biology. These neurons fuse people at a bodily level, which is why we feel unease at seeing Jackson dangle his baby, or why words come to our mouths while watching *The King's Speech.*

This discovery nicely fits the earliest accounts of human empathy, which concerned aesthetic perception. Why, for example, do we watch ballet? Would ballet be equally pleasing aesthetically if we saw rubber tires bounce around on stage in the same choreography? Would an opera be equally thrilling if the actors didn't sing, but played the banjo or accordion to serenade each other or convey their jealousy? I doubt it. While watching ballet, we enter the dancers' bodies, making every step and pirouette with them. One dancer throws another into the arms of a third, and for a second we, too, are suspended in the air. Since the audience inhabits the scene, a failed jump gets an instant reaction. If empathy were wholly cognitive, we'd expect a pause (in which the audience wonders, "How could this have happened?" or

"Is she hurt?"), but we utter "ooh's" and "aah's" even before the dancer hits the floor.

Opera creates the same connection through the human voice. Since birth (and even before) we know the voice as the vehicle of pleasure, pain, rage, and so on. The voice plugs directly into our central nervous system. It reaches inside of us as no artificial instrument ever will. We do not just infer the suffering of the soprano; we actually feel it and get goose bumps from it. As an opera lover, I feel emotionally drained by the end of every great performance.

Visual art exploits the same connection. Anyone who looks at Michelangelo's slave statue, in which a life-sized person struggles to free himself from a block of marble, feels his or her muscles straining. Standing in front of Caravaggio's painting called *The Incredulity of Saint Thomas*, in which Jesus watches his doubting disciple poke an index finger into his chest wound, we flinch at the pain this must have caused. Bosch's work, too, is full of physical scenes that arouse empathy, pity, or horror. Bosch himself has been called a misanthrope, yet his art would be nowhere without his audience's empathy. We suffer along with sinners pierced by knives, hanging half-dead in trees, eaten by hungry dogs, strung up in a harp, having a flute stuck into their anus, forced into slave labor, or ending up in a frying pan. Ironically, torture requires empathy, too, in the sense that one cannot deliberately inflict pain without realizing what is painful. That Bosch arouses as many emotions as he does is because we cannot help entering the depicted scenes. He wants us to believe that the tormented figures could be us, and for a moment they are.

Bodily empathy applies even to abstract art. An article by Vittorio Gallese, the Italian co-discoverer of mirror neurons, and David Freedberg, an American art historian, explains how we unconsciously trace the artist's movements on the canvas. In the same way that a pianist cannot listen to a piano concert without activating the motor areas in his brain responsible for finger movements, viewers of a Jackson

In Bosch's *Last Judgment*, two old hags cook humans on a spit and in a frying pan. Visualization of the sinners' ordeal is far more effective than any verbal account. We relate to human bodies at a subconscious level, literally feeling the heat of this scene.

Pollock painting experience "a sense of bodily involvement with the movements that are implied by the physical traces—in brush marks or paint drippings—of the creative actions of the producer of the work."[8]

These processes are by no means limited to our species. Amid all the academic brouhaha surrounding mirror neurons, it tends to be forgotten that they were discovered not in humans but in macaques. And still today, the evidence is better and more detailed for "monkey-see, monkey-do" neurons than for their equivalent in the human brain. Most human studies merely *assume* these neurons in certain brain areas, because to confirm their presence would require inserting electrodes, which is rarely done.[9] In monkeys, however, we have plenty of direct evidence. Mirror neurons probably help primates imitate others, such as when they open a box in the same way as a trained model or when in the wild they remove the seeds from a fruit in the way they have seen their mother do. All primates are conformists. Not only do they imitate; they also like to be on the receiving end of it. In one experiment, two investigators interacted with capuchin monkeys who were given a plastic ball to play with. One experimenter mimicked every move the monkeys made with the ball, while the other did not. By the end, the monkeys preferred the one who had imitated them. Similarly, human adolescents going out with a date instructed to mimic their every move, such as picking up a glass or leaning an elbow on the table, report liking him or her better than those going out with a nonimitating date.

It is easy to see how bodily connections aid empathy. Talking with a sad person, we adopt a sad expression and slumping body posture. We may even cry along with the other. Talking with a bouncy, giggling person, on the other hand, we soon will be laughing ourselves and feel happy as a result. The same contagion works in animals, even though this topic is understudied because of the unfortunate taboo on animal emotions. B. F. Skinner belittled emotions, particularly those of animals, saying that "the 'emotions' are excellent examples of the fictional causes to which we commonly attribute behavior."[10] Skinner's influence was immense—his school was like a religion—but is mercifully on its way out. Brain research has managed to cut the Gordian knot of skepticism about animal emotions. The amygdala in the human brain, for example, is active when we look at gory images of open wounds and violence not unlike those in Bosch's hell scenes. Electrically stimulating the same amygdala in the brain of rats causes them to flee, defecate, and crouch in a corner. It is hard to avoid the conclusion, then, that rats and humans share the same emotional state in the same part of the brain: fear. Applying this logic to love, joy, anger, and so on, modern neuroscience freely explores the emotional lives of animals.

I have never felt any attraction to the view of animals as stimulus-response machines; it is so impoverished that I don't even know where to begin taking it apart. Even Skinner himself changed his mind, as has been told by Temple Grandin, the animal expert with autism. As an eighteen-year-old student, Grandin had a meeting with Skinner, which she describes as rather awkward, including the need to explain to the professor that he shouldn't be touching her legs. She asked him whether it wouldn't be great if we knew more about the brain. Skinner answered, "We don't need to learn about the brain, we have operant conditioning." I find this absolutely shocking, because why would any scientist disdain knowledge about anything? Isn't knowledge invariably good? Unless, of course, it might threaten a pet theory! Did Skinner, like so many scientists, suffer from a disconfirmation bias? Given how much her own problems related to brain functioning, Grandin respectfully disagreed. She reports, however, that Skinner saw the light toward the end of his life when he learned firsthand that conditioning isn't everything. To a similar question about whether knowledge about the brain might be useful, he replied, "Ever since my stroke, I've thought so."[11]

Neuroscience offers two basic messages about empathy. The first is that there is no sharp dividing line between human and animal emotions. The second is that empathy runs from body to body. You stick a needle in a woman's arm, and the pain centers in her husband's brain light up just from watching the procedure. His brain reacts as if the needle went into *his own* arm. Given what we know about mirror neurons, mimicry, and emotional contagion, this "body channel" of empathy is probably as old as the primate order, but I suspect it to be much older still. Since I have devoted an entire book, *The Age of Empathy*, to documenting its origin, I will limit myself here to just a few illustrations.

One of my co-workers is replaying a famous children's song by having a hole in his bucket. Matt Campbell walks around with a plastic bucket in the bottom of which he has cut a small hole. Our chimps have learned to press an eye to the hole to watch an iPod held up at the other end of the bucket. This way, we know exactly who is watching and prevent others from seeing the videos. The whole purpose of this "peep show" is to measure yawn contagion, a peculiar phenomenon linked to empathy. For example, humans most prone to yawn contagion also have the most empathy. And children with empathy deficits, such as those with autism, lack yawn contagion altogether. Watching videos of yawning apes, our chimps yawn like crazy, but only if they personally know the ape in the video. Videos of strangers have no effect. This suggests that it is not just a matter of seeing a mouth open and close: identification with the videotaped individual is part of it. The same role of familiarity is, by the way, known of all empathy research, whether on humans or other animals. Empathic reactions are stronger the more we share with the other and the closer we feel to them. In a human field study (conducted undercover in restaurants, waiting rooms, and so on), yawn contagion was quicker and more common between relatives and close friends than between acquaintances and strangers.[12]

As an amusing aside—especially given my conversation with the Dalai Lama—I need to mention a recent IgNobel Award. This parody of the Nobel Prize, which honors research that "first makes people laugh, and then think," was given in 2011 to an attempt to find contagious yawning in turtles. Investigators at the University of Vienna exposed red-footed tortoises to a member of their own species trained to open and close its mouth. Since they observed no reaction at all, they concluded that yawn contagion is not a simple reflex but relies on mimicry and empathy, which turtles lack.

Lest one thinks that empathic mimicry in primates is limited to experimental conditions, there are plenty of spontaneous examples. Long ago, I observed at the Arnhem Zoo how an injured male chimpanzee hobbled around supporting himself on a bent wrist rather than his knuckles. Soon, the colony's juveniles hobbled single file behind this unlucky male, all leaning on their wrists. I also once witnessed a birth here at the Yerkes field station, which triggered bodily identification of one female with another:

> *From my observation window I saw a crowd gather around May—quickly and silently, as if drawn by some secret signal. Standing half upright with her legs slightly apart, May cupped an open hand underneath of her, ready to catch the baby when it would pop out. An older female, Atlanta, stood next to her in similar posture and made exactly the same hand movement, but between* her own *legs, where it served no purpose. When, after about ten minutes, the baby emerged—a healthy son—the crowd stirred. One chimpanzee screamed, and some embraced, showing how much everyone had been caught up in the process. Atlanta likely identified with May because she'd had many babies of her own. As a close friend, she groomed the new mother almost continuously in the following weeks.*[13]

Another report concerns a handicapped chimp in Budongo Forest, in Uganda. Catherine Hobaiter describes a nearly fifty-year-old male, Tinka, with severely deformed hands and paralyzed wrists. On top of this, he suffered from a chronic skin infection that caused serious discomfort, especially since he couldn't scratch himself with his atrophied limbs. Tinka, however, developed a liana-scratch technique, which resembled the way we stretch a towel between both hands to dry our back. He'd pull a hanging liana taut with his foot to rub his head and body sideways against it. It was an odd procedure, but apparently effective because he used it often. There would have been no reason for any able-bodied chimpanzees to do the same, yet several juveniles were seen to follow Tinka's lead. They regularly rubbed themselves against lianas pulled down for this purpose. Unknown from any other chimpanzee population, the spreading of this curious habit seems yet another case of mirror neurons at work.

Because of its unconscious nature, we tend to underestimate the body channel of empathy. I have even heard political commentators compare empathy to a "fragile flower," not worthy of a major role in society. Obviously, people have an agenda saying such things, while forgetting that Abraham Lincoln saw this nation as being held together by the bonds of sympathy. Lincoln's decision to fight slavery, for example, owed much to his emotional response to the misery of others. The memory of slaves shackled together with irons was a "continued torment," he wrote a friend in the South.[14] If momentous political decisions can be motivated, at least in part, by empathy, there is no reason to downplay its importance. Personally, I feel that a society without empathy and solidarity would hardly be worth living in.

Human empathy doesn't stop at the species boundary, though. It also invades policies regarding animals in our care. As an example, take the debate about the castration of pigs. In many countries, this is or was done without anesthesia. On a committee looking into this matter, opponents of the practice ran into stiff resistance from scientists and veterinarians, who asked what we know about pain and how we could possibly measure it. It was the usual platitude of "we have no idea what they feel." At the next meeting, the opponents brought a video. They didn't give any opinion except to say that since the discussion was about procedures, it would be only logical to look at current practice. They then played the scene of an awake pig being castrated. The pig made such a ruckus, squealing for minutes on end, that by the end the roomful of men looked very pale indeed, sitting with their hands firmly between their legs. More than any rational argument, this video changed the tide on anesthesia.

Such is the power of bodily empathy.

Saved by a Rat

"Who is my neighbor?" asked a lawyer puzzled by Jesus's recommendation to "love your neighbor as yourself." Finding certain people hard to love, he was looking for a less sweeping directive. The answer came in the parable of the good Samaritan.

A half-dead victim, left by the side of the road, is ignored first by a priest, then by a Levite—both were religious men familiar with the fine print of ethics. They didn't like to interrupt their journey, though, and quickly moved by on the other side of the road. Only a third passerby, a Samaritan, stopped, bandaged the man's wounds, put him on his donkey, and brought him to safety. Belonging to a class of "unclean" people despised by the Jews (for whom the parable was intended), the Samaritan was nevertheless the only one to show heart. He was affected by the plight of the wounded man at a visceral level. The biblical message is to be wary of ethics by the book, which as often as not offers excuses to ignore the plight of others.

This is only one of the parable's lessons, though. Another one is that everyone is our neighbor, even people unlike us. Given the parochialism of human and animal empathy, this is the more challenging message. Even with a simple measure, such as yawn contagion, identification with strangers is hard to demonstrate. Both chimpanzees and people join the yawns of familiar individuals more readily than those of outsiders. Empathy is hopelessly biased, as was shown, for example, in a study at the University of Zurich, which measured neural responses to the suffering of others. Men watched either a supporter of their own soccer club or a supporter of a rival club getting hurt through electrodes attached to their hands. Needless to say, the Swiss take their soccer seriously. Only their own club members activated empathy. In fact, seeing fans of the rival club getting shocked activated the brain's pleasure areas.[15] So much for loving thy neighbor!

That this in-group bias is as old as empathy itself has become clear from rodent research. Laboratory mice in transparent glass tubes could see each other while one of them received diluted acetic acid, which gave them—in the words of the investigators—a mild stomach ache. The mouse responded to this treatment with stretching movements, suggesting discomfort. The watching mouse grew more sensitive to pain, as if it experienced the other's pain itself. This experiment on commiseration worked only between mice who had lived together, however. Seeing an aching stranger left them cold.[16]

The mice showed *emotional contagion*, also well known in humans. We all know how joy spreads, or sadness, and how much we are affected by the moods of those around us. The best road to happiness, it is said, is to surround yourself with happy people. Emotional contagion has been studied by exploiting the claim that the average person fears public speaking more than death. With little preparation, subjects were asked to address an audience. Following their lecture, all participants were invited to spit into a cup. This allowed scientists to extract cortisol, a hormone associated with anxiety. They found that the speaker's stress rubbed off on his or her audience. The audience followed every word, feeling relaxed with confident speakers, but uncomfortable with nervous ones. Relying on the same body channel that was discussed earlier for *The King's Speech*, the hormone levels of speakers and audiences converged.[17]

A second rodent study involved rats. Despite the bad reputation of these animals, I have no trouble relating to its findings, having kept rats as pets during my college years. Not that they helped me become popular with the girls, but they taught me that rats are clean, smart, and affectionate. In an experiment at the University of Chicago, a rat was placed in an enclosure where it encountered a transparent container with another rat. This rat was locked up, wriggling in distress. Not only did the first rat learn how to open a little door to liberate the second, but its motivation to do so was astonishing. Faced with a choice between two containers, one with chocolate chips and another with a trapped companion, it often rescued its companion first. If the choice was between an empty container and one with chocolate, on the other hand, it invariably opened the second one first. The finding is about as contrary to the Skinnerian emphasis on conditioning as possible, and a testimony to the power of animal emotions. Interpreting the rats' behavior as empathy-based altruism, the authors concluded that "the value of freeing a trapped cagemate is on par with that of accessing chocolate chips."[18]

This jail-break experiment concerned a more complex type of empathy, known as sympathy. We don't know exactly how empathy translates into helping or comforting behavior, but it minimally requires orientation to the other. Empathy can be quite passive, reflecting mere sensitivity, whereas sympathy is outgoing. It expresses concern for others combined with an urge to ameliorate their situation. This is what the parable of the good Samaritan is all about. Unless one is made of stone, the sight of a groaning figure by the side of the road inevitably activates empathy. But instead of translating this feeling into sympathy, the two holy men tried to get rid of it. They deliberately moved away from its source. Self-protection is common, such as when people in a movie theater slam their hands over their eyes to avoid watching a gruesome scene. As a result, it is said that few audience members actually saw the key scene of *127 Hours*, a movie in which a man pinned under a boulder amputates his own arm with a pocket knife. The Samaritan, in contrast, not only faced the other's distress but showed sympathy as well. Instead of worrying about lost time, soiled clothes, or robber's tricks, he gave priority to another man in need.

The multitude of excuses humans come up with *not* to act was once tested in an ingenious experiment. University students were ordered to hurry from one campus building to the next while a slumping "victim" was planted in their path. Only 40 percent asked the "victim" what was wrong. Students who had to make haste helped far less than students with time on their hands. Some literally stepped over the

Empathy is multilayered like a Russian doll. At its core is the capacity to match another's emotional state. Around this core, evolution has built ever more elaborate capacities, such as feeling concern for others and adopting their viewpoint. Few species show all layers, but the core capacity is as ancient as the mammals.

moaning "victim." They did so even though, ironically, the topic they were to address in their lecture was the good Samaritan.[19]

The decision to help doesn't depend just on rational evaluations, however, because the driving force is almost always emotional. If it wasn't for feelings of empathy and sympathy, it is unlikely we would ever be moved to assist others. Who would dive into a river to rescue another purely on the basis of rational reflection? In the study of Swiss soccer fans, for example, the more empathy was activated in a subject's brain, the harder he tried to reduce the other's pain. On the other hand, emotions are insufficient. They combine with cost/benefit calculations to arrive at a plan of action, or inaction. This is why not all students in the good Samaritan study were willing to help. Human aid is produced by a combination of emotional drivers and cognitive filters. The same combination operates in other animals.

One way to study the role of empathy is to observe responses to distress. In chimpanzees and bonobos, consolation is a predictable outcome. A victim of aggression, who not long ago had to run for her life, now sits alone, pouting, licking an injury, or looking dejected. She perks up when a bystander comes over to give her a hug, groom her, or carefully inspect her injury. Consolations can be quite emotional, with the two apes literally screaming in each other's arms. Combing through nearly four thousand observations, we found consolation to be given mostly by friends and relatives, and more by females than males.[20] The latter also applies to our own species, in which consolation is considered a form of sympathetic concern. It is typically studied by asking family members in the home to act as if they are hurt or sad, and to see how children react. At an early age, children show the same touching, embracing, and calming body contacts as apes, and girls do so more often than boys.

I refuse to use a different terminology for these reactions in humans and apes, as urged by the opponents of anthropomorphism. Those who exclaim that "animals are not people" tend to forget that, while true, it is equally true that people are animals. To minimize the complexity of animal behavior without doing the same for human behavior erects an artificial barrier. I personally adhere to a different law of parsimony, according to which, if two closely related species act the same under similar circumstances, the mental processes behind their behavior are likely the same, too. The alternative would be to postulate that, in the short time since they diverged, both species evolved different ways of generating the same behavior. From an evolutionary standpoint, this is a convoluted proposal. Unless it can be proven that if an ape comforts another its motivation differs from that of a person doing the same, I prefer the more elegant assumption that both species follow the same urges.

The Other's Perspective

An even more complex expression of empathy is *targeted helping*. Instead of reacting to the distress of others, the goal here is to understand their situation. We recognize the specific needs of others, as when we help a blind person cross the street. We can imagine what blindness means and gear our assistance to this specific condition. There is an abundance of real-life human examples, and the same is true for other large-brained species, including dolphins, elephants, and the apes. I used to tell how a bonobo rescued a stunned bird that had flown against glass or how a chimpanzee literally dragged a wildlife-naïve friend away from a poisonous snake. There are so many stories in which one ape seems to take another's perspective. But I'll skip these stories now that targeted helping has finally been put to a test. Carried out at the Primate Research Institute (PRI) of Kyoto University, in Japan, it nicely complements our work on the willingness of chimps to do each other a favor.

I have been at the PRI several times. The chimps live in large outdoor areas with lots of green shrubs and tall climbing frames. Similarly to how we operate at Yerkes, they are called inside for voluntary testing. But at the PRI the apes have to travel through an elaborate system of tunnels before they end up in a room in which they are central and the humans peripheral. With the apes contained in a glass room, human experimenters walk around them with fancy equipment. In the experiment in question, the equipment wasn't so advanced, though. Shinya Yamamoto gave the apes a choice between two ways to obtain orange juice. They could either move a container close with a rake or suck up the juice through a straw. The problem was that they didn't have any tools available. Next to them, in a separate area, sat another chimp who had a whole set of different tools. This chimp would take one look at the other's problem, then pick out the right tool for the task and hand it to the other through a small window. If the chimp with the tools was unable to see the other's situation, however, he picked tools at random, indicating that he had no idea what the other needed. This experiment demonstrated that not only are chimps ready to assist each other but they also take the other's specific needs into account.[21]

In case anyone thinks that apes would never share tools in the field, Yamamoto's study received a nice confirmation from Fongoli, a site in Senegal where an American primatologist, Jill Pruetz, studies savanna-dwelling chimps. In contrast to forest chimps, this community has to travel enormous distances to find food. Chimps are well known to share meat, but the Fongoli chimps also share plant foods (such as baobab fruits) and are the first for which tool sharing has been reported. An adolescent female, for example, was fishing for termites with a twig when a dominant male sat down beside her with a prepared tool in his mouth. Chimps make fishing tools by breaking off a twig and cleaning it of side branches. When her own tool became ineffective, the female just took the tool out of the male's mouth, whereupon he made another one and waited beside her. This one, too, was removed from his mouth for her fishing activities. After the female left, the male made no more tools and didn't fish for insects himself.[22]

We still know little about the capacities of apes, both in captivity and in the field, but in the last few years we have been getting closer. Clearly, they are not nearly as selfish as has been assumed, and might actually beat the average priest or Levite when it comes to humane behavior.

Chapter 6

TEN COMMANDMENTS TOO MANY

> *Two things fill the mind with ever new and increasing wonder and awe, the oftener and more steadily we reflect on them: the starry heavens above and the moral law within.*
>
> —Immanuel Kant[1]

> *Can we help feeling pain when the fire burns us? Can we help sympathizing with our friends? Are these phenomena less necessary or less powerful in their consequences, because they fall within the subjective sphere of experience?*
>
> —Edward Westermarck[2]

At Tama Zoo, in Tokyo, I witnessed a surprising ritual. From the rooftop of a building, a caretaker spread handfuls of macadamia nuts among fifteen chimpanzees in an outdoor area. The macadamia is the only commercially available nut that most female chimps cannot crack with their teeth. The colony lacked adult males (I visited a small shrine with fresh flowers for Joe, their longtime alpha male, who had died a few weeks earlier), who do possess the jaw power to crack even this tough nut. The chimps rushed about collecting as many macadamias as they could in their mouths, hands, and feet. Then they sat down at separate locations in the enclosure, each with a neat little pile of nuts, all oriented toward a single place known as the "cracking station."

One chimp walked up to the station, which consisted of a big rock and a smaller metal block attached to it with a chain. She then placed a nut on the rock's surface, lifted the metal block, and hammered until the nut gave up its kernel. This female worked with a juvenile by her side, whom she allowed to profit from her efforts. Having finished her pile, she then made room for the next chimp, who placed her nuts at her feet and started the same procedure. The zookeeper explained that this was a daily ritual that always unfolded in the same orderly fashion until all nuts had been cracked.

I was struck by the scene's peacefulness, but not fooled by it. When we see a disciplined society, there is often a social hierarchy behind it. This hierarchy, which determines who can eat or mate first, is ultimately rooted in violence. If one of the lower-ranking females and her offspring had tried to claim the cracking station before their turn, things would have gotten ugly. It is not just that these apes knew their place; they knew what to expect in case of a breach of rule. A social hierarchy is a giant system of inhibitions, which is no doubt what paved the way for human morality, which is also such a system.

Impulse control is key.

The Elusive Wanton

When a Frenchwoman accused "DSK" (Dominique Strauss-Kahn, a prominent politician) of sexual assault, she couldn't resist adding that he had behaved like a "randy chimpanzee."[3] As soon as humans lose control over their impulses, we feel the need to compare them to animals. It was a terrible insult . . . to the chimpanzee!

In academic circles, too, it is impossible to avoid the popular image of out-of-control animals. This is critical in relation to moral evolution, because the opposite of morality is that we just do "what we want," the underlying assumption being that what we want is not good. An otherwise great philosopher of naturalized ethics, Philip Kitcher, once labeled chimpanzees "wantons," defined as creatures vulnerable to whichever impulse strikes them. The maliciousness and lasciviousness associated with this term was not part of his definition, which focused on a disregard of behavioral consequences. But the message was the same as that of the Frenchwoman: like some despicable men, animals lack any and all emotional control. Kitcher went on to speculate that somewhere in our evolution we overcame this wantonness, which is what made us human. This process started with the "awareness that certain forms of projected behavior might have troublesome results."[4]

Did Kitcher mean to imply that every cat who spots a mouse will blindly go after it? Would cats have no choice but to follow their hunting urge? Why, then, does she slink down with her ears pressed against her head, hide behind the trash can, and stalk the object of her desire slowly, inch by inch? Why does she waste precious minutes sneaking forward only when the mouse can't see her? Could it be that she realizes that it's better to pounce at the right moment than prematurely? I often feel like encouraging philosophers to take a pet. Learned consequences are powerful shapers of behavior.

Those zoo chimps, too, demonstrated a firm barrier between impulse and action. They obviously all wanted to crack their nuts right away, but were prevented from doing so. Or, imagine a mother chimp whose infant has been picked up by a well-meaning adolescent. The mother will follow, whimpering and begging, trying to get her offspring back from the kidnapper, who keeps evading her. The mother suppresses an all-out pursuit for fear that the adolescent will escape into a tree and endanger her precious baby. She needs to stay calm and collected. Once the infant is safely back on her belly, however, everything changes. I have seen mothers turn on the adolescent, chasing her over long distances with furious barks and screams, releasing all their pent-up frustration. Similarly, a young male not allowed to mate in view of others will stealthily sit near a sexually attractive female, making subtle signals visible only to her, spreading his legs to show off his erection and making beckoning hand gestures. He invites her to follow him to a quiet spot. One time, a young male dropped his hands over his penis as soon as an older one came around the corner, quickly hiding his intentions.

High rankers, too, benefit from impulse control. For example, an alpha male may receive a pointed challenge from a younger male, who throws rocks in his direction or makes an impressive charging display, with all his hair on end, a bit too close to the boss. This is a way of testing his nerves. Experienced alphas totally ignore the din, as if they barely notice, after which they take their time doing the rounds grooming their allies before launching a counteroffensive later in the day. By then, the young hothead will be outgunned.

How inhibited male chimps are was brought home to me when a fieldworker told me that he had never realized that male chimps could crack each other's bones. I had never thought of this either, but it makes perfect sense for an animal that can crack macadamia nuts with its teeth (which takes 300 pounds of pressure per square inch). Having recorded hundreds of encounters between members of different communities in the forest, Christophe Boesch noticed that when chimp males grab a leg and bite a stranger one can literally hear bones breaking.[5] I myself have never witnessed anything like it in fights among chimps familiar to each other, however bad these fights might seem. This means that most of the time, at least within their group, male chimps restrain their violent potential.

The beauty of an emotional response system over an instinctual one is that its outcome isn't set in stone. The term "instinct" refers to a genetic program that tells animals, or humans, to act in a specific way under specific circumstances. Emotions, on the other hand, produce internal changes along with an evaluation of the situation and a weighing of options. It is unclear whether humans and other primates have instincts in the strict sense, but there is no doubt about their having emotions. Klaus Scherer, a German expert, calls emotions "an intelligent interface that mediates between input and output on the basis of what is most important to the organism at a particular time."[6]

This may seem counterintuitive since it depicts emotions as intelligent, but keep in mind that the whole distinction between emotion and cognition is under debate. The two are intertwined. Furthermore, their interplay is probably very similar in humans and other primates. The prefrontal cortex, which helps regulate emotions, is often assumed to be exceptionally large in our own species, but this is an outdated view. The human cerebral cortex holds 19 percent of all neurons in the brain, just like any typical mammalian brain. For this reason, our brain has been called a "linearly scaled-up primate brain." It may be large overall, but the way its various parts relate to each other is unexceptional.[7]

Most of us have seen the hilarious videos of children sitting alone at a table desperately trying *not* to eat a marshmallow—secretly licking it, taking tiny bites from it, or looking the other way so as to avoid temptation. It is one of the most explicit tests of impulse control. The children have been promised a second marshmallow if they leave the first one alone. Such "deferred gratification" has also been tested in our primate relatives. For example, monkeys will leave a slice of banana alone if they know that by doing so they may get a larger one later on. Or, a chimp patiently stares at a container into which falls a candy every thirty seconds. At any moment, he can disconnect the container and swallow its contents, but then the candy flow will stop. The longer he waits, the more candies he will get. Apes do about as well as children on this task, delaying gratification for up to eighteen minutes. They wait longer if they have toys to take their mind off the candy machine. Like children, they seek distraction to better fight temptation. Does this mean that they're aware of their own desires, and deliberately curtail them? If so, we seem to be getting rather close to free will![8]

Clearly, Kitcher's "wantons" are a nonexistent species. Primates offer great insight into group life based on both emotions and emotional control. Tightly embedded in society, they respect the limits it puts on their behavior and are ready to rock the boat only if they can get away with it or if so much is at stake that it's worth the risk. Otherwise, like the chimpanzees at Tama Zoo, they await their turn and control their urges. We come from a long line of ancestors with well-developed hierarchies for whom social inhibition was second nature. If doubters need proof of how much we owe to this history, they need only consider how much we invest moral rules with authority. Sometimes the authority is personal, like a super alpha male, as when we claim that God handed us the rules on a mountaintop. At other times, we fall for the authority of reasoning, claiming that certain rules are so logically compelling that it would be silly to disobey them. Humanity's reverence for the moral law betrays the mindset of a species that likes to stay on good terms with higher-ups.

Nothing is more telling than how we react after a transgression. We lower our face, avoid the gaze of others, slump our shoulders, bend our knees, and generally look diminished in stature. Our mouth droops and our eyebrows arch outward in a distinctly unthreatening expression. We feel ashamed and hide our face behind our hands or "want to sink into the ground." This desire for invisibility is reminiscent of submissive displays. Chimpanzees crawl in the dust for their leader, lower their body so as to look up at him or turn their rump toward him to appear unthreatening. Dominant apes, in contrast, make themselves look larger and literally run or walk over a subordinate, who ducks

A dominant chimp, his hair bristling, walks bipedally while carrying a big rock. He looks larger than his rival, who evades him with pant grunts, a sign of submission. It all comes down to ritual, however, because in reality these two males are of equal weight and size.

into a fetal position. Daniel Fessler, an anthropologist who has studied shame in human cultures, compares its universal shrinking appearance to that of a subordinate facing an angry dominant. Shame reflects awareness that one has upset others, who need to be appeased. Whatever self-conscious feelings go with it, they are secondary to the much older hierarchical template.

The only uniquely human expression, as Darwin already noted, is blushing. I don't know of any instant face reddening in other primates. Blushing is an evolutionary mystery that must be particularly perplexing for those who believe that exploitation of others is all that humans are capable of. If this were true, wouldn't we be better off without blood uncontrollably rushing to our cheeks and neck, where the change in skin color stands out like a light tower? Such a signal makes no sense for a born manipulator. The only advantage of blushing that I can imagine is that it tells others that you are aware how your actions affect them. This fosters trust. We prefer people whose emotions we can read from their faces over those who never show the slightest hint of shame or guilt. That we evolved an honest signal to communicate unease about rule violations says something profound about our species.

Blushing is part of the same evolutionary package that gave us morality.

One-on-One Morality

Morality is a system of rules concerning the two *H*'s of Helping or at least not Hurting fellow human beings. It addresses the well-being of others and puts the community before the individual. It does not deny self-interest, yet curbs its pursuit so as to promote a cooperative society.

This functional definition sets morality apart from customs and habits, such as eating with knife and fork versus chopsticks or one's bare hands. People may disapprove of my eating with my hands, at least in my current culture, but their disapproval is not of a moral nature. Even young children distinguish etiquette (boys go the boys' toilet and girls to the girls' toilet) from moral rules (don't pull at ponytails). Rules related to the two *H*'s are taken far more seriously than conventions. Toddlers believe in the universality of the former. If you ask them whether they can imagine a culture where everyone goes to the same toilet, they can, but ask them whether there may be cultures where it is perfectly fine to hurt someone else unnecessarily, they refuse to believe so. As the philosopher Jesse Prinz has explained, "Moral rules are directly grounded in the emotions. When we think about hitting, it makes us feel bad, and we cannot simply turn that feeling off."[9] Moral understanding develops astonishingly early in life. Infants under one year of age already favor the good guy in a puppet show. The puppet who nicely rolls a ball back and forth with another is preferred over one who steals the ball and runs off with it.

Empathy is critical. At a young age, the child learns that slapping or biting a sibling produces a screaming reaction with a range of negative consequences. Apart from those few children who will grow into psychopaths—whose childhood is marked by animal torture, excessive violence, and lack of remorse—the vast majority doesn't enjoy the sight of a crying sibling. Second, hurting your sibling brings all fun and games to an abrupt halt. No one wants to play with someone who lashes out all the time. Finally, an angry parent or teacher is likely to enter the scene and yell at the hitter or make him feel guilty by pointing out his victim's tears. All of these emotional consequences discourage hurting a playmate. Empathy teaches children to take the feelings of others seriously.

Not that we should be surprised by such early development. It is only because of the prevalence of Veneer Theory that it was believed that goodness is not part of human nature, and that we need to work hard to teach it to our children. Children were seen as selfish monsters, who learn to be moral from teachers and parents *despite* their natural inclinations. They were seen as reluctant moralists. My view is the exact opposite. The child is a natural moralist, who gets a huge helping hand from its biological makeup. We humans automatically pay attention to others, are attracted to them, and make their situation our own. Like all primates, we are emotionally affected by others. And not just like primates. The reason a big dog will stop gnawing on a smaller playmate as soon as the latter utters a sharp yelp is the same: hurting another is aversive.

Dogs are more hierarchical than us, so have additional reasons to fear consequences. This may explain the strong reaction of Bully, a dog owned by Konrad Lorenz, to a rule violation. In this case, the victim was not a vulnerable other but the master himself. Bully accidentally bit the famous ethologist's hand when the latter tried to break up a fight. Even though Lorenz did not reprimand him and immediately reassured him, Bully suffered a complete nervous breakdown. He was virtually paralyzed for days and uninterested in food. He would lie on the rug breathing shallowly, occasionally interrupted by a deep sigh coming from deep inside his tormented soul. He looked as if he had come down with a deadly disease. Bully remained subdued for weeks, prompting Lorenz to speculate about his having a "conscience." Bully had never bitten any person before, and so he could not have relied on previous experience to decide that he had done something wrong. Perhaps he had violated a natural taboo on inflicting damage upon a superior, which could have the worst imaginable consequences, including expulsion from the pack.[10]

As a student, I followed the behavior of a group of macaques in the presence or absence of the alpha male. As soon as alpha turned his back, other males began to approach females. Normally, they'd get into trouble for doing so. When this principle was put to the test, low-ranking males refused to approach females so long as the dominant looked on from inside a transparent box, yet dropped all inhibitions as soon as he was removed. They felt free to copulate. The same males also performed bouncing displays and walked around with their tails in the air, which is typical of high-status males. Upon the boss's return, however, they were excessively nervous, greeting him with wide submissive grins on their faces. They seemed to realize they had done something wrong.[11]

When-the-cat-is-away situations are always amusing to watch since the cat is never far from the mice's minds. In another macaque group, a longtime alpha male, Mr. Spickles, would sometimes get tired of keeping an eye on half a dozen restless males during the breeding season. Or, perhaps he just wanted to warm his old bones indoors. He would disappear for half an hour at a stretch, leaving the others plenty of opportunities to mate. The beta male was very popular with the ladies, but so obsessed with Mr. Spickles that he was irresistibly drawn to the door to peek inside through a crack. Perhaps he wanted to make sure Mr. Spickles stayed where he was. Rhesus monkeys being multiple-mount-ejaculators (requiring several mounts before the male ejaculates), the young male would race back and forth a dozen times between his partner and the door before completing a mount series.

Social rules are not simply obeyed in the presence of dominants and forgotten in their absence. If this were true, low-ranking males wouldn't need to actively check on alpha or be overly submissive following their exploits. They internalize rules to some degree. A more complex expression occurred once in the Arnhem chimpanzee colony following the first time the beta male, Luit, had physically defeated Yeroen, the alpha. The fight had occurred while both males were alone together in their night quarters. The next morning, the colony was released onto its island only to notice the shocking evidence:

> *When Mama discovered Yeroen's wounds she began to hoot and look around in every direction. At this, Yeroen broke down, screaming and yelping, whereupon all the other apes came over to see what was the matter. While the apes were crowding around him and hooting, the "culprit," Luit, also began to scream. He ran nervously from one female to the next, embraced them, and presented himself to them. He then spent a large part of the day tending Yeroen's wounds. Yeroen had a gash in his foot and two wounds in his side, caused by Luit's powerful canines.*[12]

Luit's situation resembled that of Bully, the dog, in that he had broken the spell of the hierarchy by biting alpha. What a terrible thing to do, the group's reaction seemed to convey, and Luit did his best to make amends. But not by giving up his strategy to dominate Yeroen, because in the weeks that followed he kept the pressure on, and in the end forced Yeroen into retirement from the top spot. His grooming of the old male's injuries shouldn't surprise anyone, for it is part of normal chimpanzee relations, but such behavior is even more typical of bonobos. During reunions among adversaries, I have seen aggressors unhesitatingly pick up the foot or reach for the arm of the other where they had placed their teeth. This suggested not only a precise recall of the fight but also regret. Biting is so rare in bonobos that it makes sense for them to be concerned about its consequences, licking up every drop of blood spilled as a result of their intemperate behavior.

Empathy probably played a role in these incidents, and also in Luit's reaction to Yeroen's injuries. Even though superficial by chimpanzee standards, they were the first gashes Yeroen had suffered in years. In general, primates are keen on keeping good relations even in the face of discord and strife. That they are well aware of the undermining effects of pain and distress is also visible in their play. When youngsters are far apart in age, games often get too rough for the younger partner, as when its leg gets twisted or a gnaw turns into a bite. At the slightest peep of distress, its mother will break up the game. Normally, play is entirely silent except for the hoarse panting laughs that sound like human laughter. Recording hundreds of wrestling bouts, we found that juveniles laugh especially when the mother of a younger playmate is watching. They laugh more in her presence than while alone with the same infant. It's almost as if they seek to reassure the mother: "Look how much fun we're having!"[13]

In sum, two great reinforcers support the social code by which primates and children live. One comes from within and the other from without. The first is empathy and a desire for good relations leading to the avoidance of unnecessary distress. The second is the threat of physical consequences, such as penalties meted out by higher-ups. Over time, these two reinforcers create an internalized set of guidelines, which I will call *one-on-one morality.* This kind of morality permits partners of disparate abilities and strength to get along, such as males with females and adults with juveniles, binding them together in a mutually agreeable modus vivendi. Sometimes these guidelines are suspended—for instance, when two rivals compete over status—but generally primates strive for peaceful coexistence. Individuals unable or unwilling to abide by the social code become marginalized. The ultimate driver of the whole process, in an evolutionary sense, is the desire for integration, since its opposite—isolation or ostracism—drastically diminishes an individual's chances at survival.

In her 1985 book *Sex and Friendship in Baboons*, Barbara Smuts was the first to apply the term "friendship" to animals, a term fiercely contested at the time. Some considered it overly anthropomorphic. But with increasing knowledge of the bonds between animals, the skepticism dissipated, and the term became commonplace. Two old male chimpanzees in Kibale Forest, in Uganda, for example, traveled together most of the time, hunting side by side, sharing meat, and calling to each other to stay in touch whenever they were separated by thick foliage. These males also backed each other in fights against third parties. They were trusted partners for many years despite being unrelated. John Mitani, a primatologist who followed them for years, relates how upon the death of one of them, his buddy became far less sociable, isolated himself, and seemed to go into mourning. Many such relationships have been documented, and DNA analysis supports the claim that they often are between nonrelatives. The friendship label is no exaggeration, therefore, and is also applied to bonds among elephants, dolphins, and other animals. Field studies on baboons have shown that females with friendships outlive those without them, and raise more offspring. There are excellent evolutionary reasons, therefore, to value close relationships.[14]

The social code to stay on good terms with everyone covers, among other things, who can mate with whom, how to play with infants, whom to defer to, and under what circumstances to appropriate another's food or await your turn. Chimpanzees and bonobos respect each other's possessions, so that even the top male may have to beg for his food. It is rare for dominant individuals to take another's food by force, and code violators meet with fierce resistance. The community attack on Volker, described in chapter 3, illustrates how wild bonobos handle those who break the rules. There is no shortage of similar examples, such as the following scene in the chimp colony under my observation window. Jimoh, the previous alpha male, once punished a younger male for a suspected mating. Normally, Jimoh would merely chase the culprit off, but for some reason—perhaps because the same female had refused to mate with himself that day—he went full tilt after his rival and did not relent. The young male had diarrhea for fear, and it looked unlikely that things would end well. Before Jimoh could get his hands on him, however, females gave loud "woaow" barks in protest. This swelled into a deafening chorus when the alpha female joined in. Once the protest reached a climax, Jimoh broke off his attack with a nervous grin on his face: he got the message.

I felt I was witnessing public opinion at work.

When "Is" Meets "Ought"

What is so fascinating about the social code is that it is prescriptive. The code has teeth. I'm not just talking here about how animals behave, but how they are *expected* to behave. It all boils down to the distinction between "is" and "ought." This may sound like an odd grammatical aside, but the is/ought divide happens to be a major topic for philosophers. In fact, it is impossible to discuss the origin of morality without getting embroiled in this distinction. The "is" describes how things are (social tendencies, mental capacities, neural processes), while the "ought" relates to how we want things to be and how we are supposed to behave. The "is" is about facts and the "ought" about values. Animals living by a prescriptive code have made a transition from "is" to "ought." And they have done so, I might add, blithely unaware of the ocean of academic ink spilt over this particular transition.

The Scottish philosopher who gave us the is/ought distinction, David Hume, wrote almost three centuries ago that we should be careful not to assume that the two are the same, adding that "a reason should be given" for how we argue from the facts of life to the values we strive for.[15] In other words, morality is not simply a reflection of human nature. Just as one cannot infer traffic rules from the description of a car, one cannot infer moral norms from knowing who or what we are. Hume's point is well taken, but a far cry from the exaggeration by later philosophers, who turned his appeal for caution into "Hume's guillotine," claiming an unbridgeable chasm between "is" and "ought." They went on to wield this guillotine to kill off any and all attempts, even the most cautious ones, to apply evolutionary logic or neuroscience to human morality. Science cannot tell us how to construe morality, they said. While that is true, science surely can help us explain why certain outcomes might be favored over others, hence why morality is the way it is. For one thing, there would be no point in designing moral rules that are impossible to follow, just as there would be no point in making traffic rules that cars can't obey, such as ordering them to jump over slower cars. Among philosophers, this is known as the "ought implies can" argument. Morality needs to fit the species it is meant for.

"Is" and "ought" are like the yin and yang of morality. We have both, we need both, they are not the same, yet they are also not totally separate. They complement each other. Hume himself ignored the "guillotine" named after him by stressing how much human nature matters: he saw morality as a product of the emotions. Empathy (which he called sympathy) was at the top of his list. He considered it of immense moral value. This opinion represented no contradiction on his part, since all he urged was caution in moving from how we are to how we ought to behave. He never said that such a move was prohibited. We should also keep in mind that the tension between the two is felt much less clearly in real life than at the conceptual level at which most philosophers like to dwell. They feel that we cannot *reason* ourselves from one level to the other, and they are right, but who says that morality is or needs to be rationally constructed? What if it is grounded in emotional values, as Hume thought?

Values are embedded in the way we are. It is sometimes thought that biology falls entirely on the "is" side of the moral equation, but every organism pursues goals. Survival is one, reproduction is another, but there are also more immediate goals like keeping rivals out of one's territory or avoiding extreme temperatures. Animals "ought" to feed themselves, escape predators, find mates, and so on. While having a full belly is obviously not a moral value, the distinction becomes harder when we get to the social domain. Social animals "ought" to get along. Human morality develops out of sensitivity to others and out of the realization that in order to reap the benefits of group life we need to compromise and be considerate of others.

Not all animals share this sensitivity. Even if they were as smart as us, piranhas or sharks would never acquire our social code, given that hurting others—except for the risk of retaliation—doesn't bother them in the least. Emotionally, we are radically different, which explains why we assign the two *H*'s of helping or (not) hurting special status. Rather than reaching us from the outside or through logic, these values are deeply embedded in our brainstem. In her book *Braintrust*, Patricia Churchland inserts is/ought language to explain how evolution predisposed us for morality:

> *From a biological point of view, basic emotions are Mother Nature's way of orienting us to do what we prudently ought. The social emotions are a way of getting us to do what we socially ought, and the reward/punishment system is a way of learning to use past experiences to improve our performance in both domains.*[16]

The tension between what we are and what we should be leads to intriguing debates, such as one I had with a blogger who claimed that people with a natural altruistic drive are less deserving of our esteem than those who have no such drive, yet still show altruism. One of the most influential moral philosophers, Immanuel Kant, thought this way. He saw about as much worth in human kindness as Dick Cheney saw in energy conservation. Cheney mocked conservation as "virtuous" yet irrelevant, whereas Kant called compassion "beautiful" yet without moral use. Who needs tender feelings if duty is all that matters?

Another blogger, in contrast, preferred people with a spontaneous impulse to help to those who help only on the basis of a calculation of what is the right thing to do. He thus favored felt kindness over altruism out of obligation. It is an interesting dilemma, comparable to the question whom you'd wish to marry: someone who loves you or someone who gives exactly the same support but only because she considers it her duty? The latter spouse is surely putting in more effort, and deserves much admiration, but I'd rather be married to the former. I may be hopelessly romantic, but I don't expect a solid a commitment out of duty. In the same way, morality will be far more reliable if genuine prosocial feelings constitute its driving force.

Hell on Earth

The second driving force is our hierarchical nature and fear of punishment. This is an old theme, especially for those who claim that morality can't exist without God. We needn't agree with their gloomy assessment, but there is no denying the role of authority and social pressure. As opposed to prosocial tendencies, which play a role early in life, rule enforcement appears rather late. Even the notoriously hierarchical rhesus macaque shows incredible tolerance of the young. In the 1980s, I conducted experiments by depriving a large monkey troop of water for a few hours, after which I'd fill up their basin. All adults came to drink in hierarchical order, not unlike the nut-cracking scene at Tama Zoo, but infants under one year of age showed up whenever they wanted. They drank with the highest-ranking males and freely mingled with the top matriarch's family. Punishment came into play only in their second year, when they quickly learned their place.

Since apes develop more slowly than monkeys, youngsters go virtually unpunished for the first four years of life. They can do nothing wrong, such as using the back of a dominant male as a trampoline, pulling food out of the hands of others, or hitting an older juvenile as hard as they can. Not even their own mother will correct them. Her main strategy is distraction. If her infant crawls toward an adult in a foul mood or is close to fighting with a playmate, the mother will tickle him, lead him away, or pick him up for some nursing. One can imagine the shock when a youngster is rejected or punished for the first time. The most dramatic punishments are those of young males who have ventured too close to a sexually attractive female. Until this time, they were allowed to hang around these females and even mate with them insofar as they were capable. Inevitably, however, the day would come that the competitive atmosphere among adult males would spill over in an impromptu attack on a youngster. One of the males, with all his hair on end, would storm at the unsuspecting Don Juan, grab him, and hurl him wildly about with his foot in his mouth, drawing blood. Young males need only one or two such lessons. From then on, every adult male can make them jump away from a female by a mere glance or a step forward.

Youngsters thus learn to control their sexual urges, or at least to become circumspect about them. Human children learn social rules in the same way. We are far more accepting of a three-year-old, whose misbehavior mostly amuses us, than of a teenager, whose transgressions deeply upset us. The learning process is the same as in other primates, from anything goes to an ever narrower range of acceptable behavior. No wonder that punishment figures prominently in our moral systems, from law enforcement to the shunning of those who have cheated us, and from an "eye for an eye" to the eternal burning in hell that awaits sinners.

To instill fear of punishment is no minor task, and both religion and society work hard on it. This is where Bosch's paintings occupy a special place. If Bosch is known for anything, it is as the painter of hell, reminding us what terrible things will befall those attracted to vice. His scenes of torture and annihilation play on our deepest fears of rejection, suffering, and death. It isn't surprising that his paintings were copied over and over, the way images are nowadays spread over the Internet. An entire art factory in Antwerp copied Bosch scenes. Yet, those who believe that Bosch must have been a deeply religious man to paint all those images should look more closely. *The Garden*'s right-hand panel was unparalleled in that Bosch omitted God, leaving everything in human hands. In the same way that Buddhism lacks a punitive God, but does have the concept of "Karma" for payback to those who lead immoral lives, Bosch painted literally a hell on earth. It is filled with so many day-to-day scenes, even though bizarre ones, that it looks like a most unpleasant way to end one's life rather than the traditional fiery pit of the afterlife. We can see fire on the horizon, but it is earth itself that is burning. Bosch's hell even includes a frozen lake with naked humans and fantasy animals skating on it, the way the Dutch skate on anything that freezes over. Certainly not your typical inferno!

An important figure is a bird-headed monster, known as the Prince of Hell, with a cauldron on his head. The monster sits on a throne-like toilet, feasting on damned humans, which he defecates into a transparent bag or amniotic sac hanging underneath. We also see two giant disembodied ears with a knife between them, the interpretation of which ranges from the knife being an alchemist's tool of purification to the ears symbolizing humanity's deafness to the New Testament, to the whole set symbolizing a canon on wheels or a phallus with a scrotum. Combined with the many outsized musical instrument in Bosch's hell (including the first accurate depiction of the hurdy-gurdy), the large ears also hint at musical torture by means of a never-ending cacophony.

Bosch leaves us mortals alone with our fate, our terror. Trying to connect this earthly hell with other elements of his triptych, I was struck by the abundance of fruits in the other two panels and their absence in hell. I have already mentioned that Bosch's paradise lacks forbidden fruits, suggesting that Adam and Eve never went through the trauma of gaining forbidden knowledge and its dire consequences. He made up for this absence by giving the masses of nudes in the central panel more fruits than they can handle. They are fed fruits by birds, feed each other, and carry giant strawberries around, including a man whose head has turned into a grape, a possible reference to a Middle Dutch expression about the glans penis.[17] Whereas the central panel is full of unforbidden fruits, however, hell is devoid of them, so one gets the impression that Bosch was making a point.

Foolishly, I thought the answer might be found in a novel entitled *Big Sur and the Oranges of Hieronymus Bosch*. I took it along on a trip to California. Sitting among nude bathers in a sulfurous spring did connect me with *The Garden*, and Big Sur was obviously magnificent, but I had great trouble getting through the book. I have rarely read a more self-indulgent author than Henry Miller, and quickly discovered that he knew next to nothing about Bosch. Miller's book mentions Bosch's triptych *The Millennium*, which is the new name a German art historian, Wilhelm Fränger, had bestowed upon *The Garden*.[18] Fränger claimed that Bosch belonged to a heretical sect, an oft-repeated speculation despite the utter lack of support for it. It belongs with other fantasy claims, such as that Bosch was a closet homosexual afraid of castration or a raging schizophrenic. Miller is enthusiastic about the depiction of an orange grove, noticing how realistic Bosch's oranges appear ("far more delectable, far more potent, than the Sunkist oranges we daily consume").[19] It is unlikely, though, that Bosch even knew what an orange was. Northern Europeans learned about this fruit only in the sixteenth century and started growing it in "orangeries" in the seventeenth century. Dutch painters, most famously Pieter Mondrian, typically feature apple trees or other northern fruits. *The Garden*, too, seems to show an apple orchard.

The skewed fruit distribution across the triptych is easily explained: fruits symbolize pleasure, both gustatory and erotic, and hell is a place where all pleasure is extinguished. This still leaves the absence of God, however. Since God is present in Bosch's other hell painting, *The Last Judgment*, where he prominently presides over the anguished masses, there must be a reason why *The Garden* leaves him out. Did Bosch have a secular message? Was he suggesting that being immoral was deserving of hellish punishment independent of divine judgment? Was he hinting at Socrates's famous question whether we need the gods to tell us what is moral? Is an action moral because the gods love it, or do the gods love an action because it is moral? Socrates asked Euthyphro.

The Garden invites us to imagine a world in which we carry out our daily business without divine instructions about right and wrong and without God's oversight. Such a world would still require morality, Bosch seems to be saying, and would still punish those who fail to

The apocalyptic right-hand panel of Bosch's *Garden* includes two ears pierced by an arrow. The ears carry a sharp knife between them. Riding roughshod over doomed souls, the ears have baffled generations of art critics.

lead decent lives even if, instead of going to hell, they'd be visited by it on earth.

Community Concern

A child of the Renaissance, Bosch lived in a time that came to value reason over piety. Humanity began to dream of a rationally justified morality, culminating centuries later in Kant's elevation of "pure reason" to its foundation. The prevailing approach was that eternally valid moral truths were somewhere "out there," held together by a compelling logic that is ours to uncover. Philosophers offered their expertise to do so.

How did such an odd idea arise? It recalls the argument from design in the evolution debate, which takes the eye as example. The eye's complex functionality could never have arisen by chance, so the argument goes, so we must assume an intelligent designer. Most biologists disagree, and point at intermediate stages, from the light-sensitive pigment spots of the flatworm to the "pinhole" eye of the nautilus. Given enough time, natural selection can produce enormously complex structures in small incremental steps, acting like a "blind watchmaker," in Dawkins's felicitous phrase. No plan is needed. So, why treat the moral law like another eye? True, it is complex and sophisticated, but that doesn't imply a logical design. What else in nature has one? The idea that morality can be argued from first principles is a creationist myth, and a poorly supported one at that, since no one has done so in any convincing manner. All we have are approximations.

Normative ethics carries the stamp from a previous era. The whole idea of a moral "law" suggests an enforced or enforceable principle, which makes one wonder who the enforcer might be. In the past the answer was obvious, but how to apply this idea without invoking God? For a philosophical take on this issue, I recommend Philip Kitcher's *The Ethical Project*, which expresses skepticism:

Theorizing about the ethical project has been hampered by assuming there must be some authority in ethics, some point of view from which truth can be reliably discerned. Philosophers have cast themselves as enlightened replacements for the religious teachers who previously pretended to insight. But why? Ethics may simply be something we work out together.[20]

Although we live in an age that celebrates the cerebral and looks down upon emotions as mushy and messy, it is impossible to get around the basic needs, desires, and obsessions of our species. Made of flesh and blood, we are driven to pursue certain goals—food, sex, and security foremost among them. This makes the whole notion of "pure reason" seem like pure fiction. Did you hear about the study showing that court judges are more lenient after lunch than before lunch?[21] For me, this is human reasoning in a nutshell. It is virtually impossible to disentangle rational decision making from mental predispositions, subconscious values, emotions, and the digestive system. According to cognitive science, rationalizations are mostly post hoc. We have a dual mentality that immediately suggests intuitive solutions, well before we've thought about the issue at hand, followed by a second, slower process that vets these solutions for quality and feasibility. While the second process helps us justify decisions, it is a gross sleight of hand to present these justifications as the actual reasons. We do this all the time, however, like the slave owner who says he's doing slaves a favor, or the warmonger who says all he wanted to do is free the world of a tyrant. We are good at finding reasons that suit our purposes. Jonathan Haidt exposed this tendency in moral arguments, aptly comparing it to the tail wagging the dog. The grounds we offer for our behavior poorly reflect actual motives. In Pascal's elegant phrasing: "The heart has its reasons, of which reason knows nothing."[22]

I am in fact so skeptical of the explanations people offer for their behavior that I feel immensely lucky to work with subjects unable to fill out questionnaires. Yet, the prevailing opinion remains that thinking precedes behavior. I have heard philosophers talk of the "idea" of forgiveness and the "concept" of fairness, even claiming we owe the latter to the French Revolution. Are they implying that before Marie-Antoinette lost her head, humans were oblivious to fairness? While we are great at translating preexisting tendencies into concepts, this doesn't prevent primates and young children, who have never heard of them, from kissing and embracing after a fight or vociferously objecting to unequal rewards or Christmas presents. Let me therefore return to my bottom-up account, which puts emotions in the driver's seat. It assumes two basic levels to morality, one regarding social relationships and the other regarding the community. The first level is what I have called one-on-one morality, which reflects an understanding of how one's own behavior affects others. We share this level with other social animals, which develop similar inhibitions and codes of conduct. Failure spells disharmony, which is why we feel an obligation, an "ought," to consider the interests of others. Reasoning is *not* at its root, even though it is not hard to come up with reasons why others would object to poor treatment or why a male chimp would punish a younger rival hanging around a female. The reactions are entirely emotional, however, such as one male's jealousy at another, or the appreciation that in order to enjoy the company of friends you'd better act like one.

One-on-one morality is rather narrow, though. We need a second level, which I call *community concern*. It doesn't deny personal interests, but it is a radical step up in that the goal is harmony within the larger community. This is the level at which human morality begins to depart from anything else encountered thus far, even though some animals show rudimentary forms of community concern.

I have already mentioned the law-and-order role of Phineas and other high-ranking primates, who break up fights among others. The same impartial "policing" is known from wild chimps, and a recent study comparing it across various groups concluded that it stabilizes social dynamics.[23] There are also the intercessions by older females, who bring warring males together, literally tugging at their arm to get them to approach their adversary. Females pry heavy rocks from violent males' hands. They do so even if they are not directly involved themselves and could easily have stayed on the sidelines. Chimpanzees thus ameliorate the social atmosphere around them, promoting peace not just for themselves but for everyone else as well. Or take the way females react to attempts by males to sexually force themselves onto one among them. In nature, males may succeed, because they can distance themselves from the group by taking a female on "safari" to avoid third-party intrusions. They may even wield branches as weapons to bully reluctant females into intercourse. In captivity, however, it is impossible to get away from the rest, and I have often seen how males whose sexual advances were too insistent triggered a screaming protest by the female in question, upon which a mass of females would help her chase off the offender. Since female solidarity is not a general rule in chimpanzees, their unanimous objection to attempted rape is remarkable. Have they reached a tacit agreement? Do females realize that if everyone assists anyone in need, all of them will be better off in the long run?

Human community concern is driven by enlightened self-interest. We strive for a well-functioning whole because this is what we thrive in. If I see a burglar breaking into a house in my street, even though it is not my house and none of my business, I will follow the rules of society and call the police. If something equivalent were to happen in an ancestral settlement, we'd mobilize everyone to stop the person who has trouble telling mine from thine. Moral transgressions, even those that don't directly affect us, are bad for everyone. There are great descriptions in the anthropological literature of how this operates in preliterate societies, such as Colin Turnbull's story of a Mbuti pygmy hunter, Cephu, who had set up his net in front of that of other families. Mbuti hunters stretch long nets in the jungle, after which women and children drive animals, like forest antelopes and forest pigs, toward the nets, making lots of noise. The trapped animals are then speared by the hunters. By placing his net ahead of the rest, Cephu made an exceptionally good catch, but unfortunately for him, the others had noticed his cheating ways. Upon return to camp, the mood was somber, and negative remarks about Cephu began to trickle in. According to Turnbull, Cephu's infraction was truly outrageous in the eyes of the usually gentle pygmies. At camp, the other hunters began ridiculing and mocking Cephu. Younger men refused to get up to offer him a seat, while others told him they'd love to see him fall on his spear. Cephu burst out in tears and soon saw all his meat being distributed to others, even the meat his wife had tried to hide under the roof of their hut. Cephu learned an important lesson, and the community enforced a rule that is the lifesaver of all hunter-gatherers. Cooperation guarantees a steady food supply for everyone, so individual hunting success needs to be downplayed and sharing to be made a deeply felt obligation.

I think of such events each time I see chimpanzee "public opinion" at work, such as when females confront an overly aggressive male. Do they oversee the supra-individual consequences, the way the Mbuti surely do? And do individuals who take the lead gain prestige from doing so? Prestige and reputation are a critical part of why humans act morally even when they don't directly gain from it. Others are more willing to follow the lead of an upstanding citizen than of someone who lies, cheats, and always puts his own interests first. Glimmers of reputation can be seen in the apes. For example, if a major fight gets out of control, bystanders may wake up the alpha male, poking him in the side. Known as the most effective arbitrator, he's expected to step in. Apes also pay attention to how one individual treats another, as in one experiment in which they preferred to interact with a human who had been nice to others. This was not about how they themselves had been treated but about the reputation the human had gained by sharing food with other apes.[24] In our own studies, we found that if we let the colony watch two chimpanzees who demonstrate different but equally simple tricks to get rewards, they prefer to follow the higher-status model. Like teenagers copying the hairstyle of Justin Bieber, they imitate prominent members of their community rather than bottom rankers.[25] Anthropologists call this the *prestige effect.*

But despite the various hints of individual reputation and sensitivity to communitywide issues in apes, humans go well beyond this. We are far better at calculating how our own actions and those of others impact the common good, and at debating among ourselves which rules to implement and what kind of sanctions to apply. We realize that even a small infraction needs to be nipped in the bud, lest the same individual proceed to more serious ones. We also have the advantage of language, which allows us to relay events that occurred distantly in time and place, so that the whole community knows about them. If one chimp mistreats another, his victim may be the only one to know. In humans, everyone will know the colorful details the next morning, including people in surrounding villages. We are incredible gossipers! Language allows us to keep memories alive and bring up certain infractions time and again. Our reputations are cumulative, stored in collective memory. Cephu's infraction will not be forgotten during his lifetime, and even his children may be reminded of it. Humans have taken reputation building and community concern to a level unmatched by anything seen in the apes, thus tightening the moral net around each individual.

Community-level thinking also explains our interest in generalizable rules. One way in which the moral emotions differ from ordinary ones is "by their disinterestedness, apparent impartiality, and flavor of generality," as Edward Westermarck put it.[26] Emotions such as gratitude and resentment concern only our personal interests—how we have been treated or how we want to be treated—whereas moral emotions go beyond this. They deal with right and wrong at a more abstract level. It is only when we make judgments of how *anyone* under the circumstances ought to be treated that we speak of moral judgment. To get the same point across, Adam Smith asked us to imagine what an "impartial spectator" would think of our behavior. What would be the opinion of someone who was not involved? This is human morality at its most complex: an opinion about right or wrong regardless of what's in it for ourselves.

I have trouble with the impartiality of the impartial spectator, however. After all, he is human and either belongs to our community or can at least imagine that he does. Smith never proposed that he was an extraterrestrial. To get some perspective on this, take a story related by Westermarck in one of his books on Morocco. It was about a vengeful camel that had been excessively beaten on multiple occasions by a fourteen-year-old "lad" for loitering or turning the wrong way. The camel passively took the punishment, but a few days later, while finding itself unladen alone on the road with the same conductor "seized the unlucky boy's head in its monstrous mouth, and lifting him up in the air flung him down again on the earth with the upper part of the skull completely torn off, and his brains scattered on the ground."[27] This horrible scene lends itself to moral interpretation, especially given the boy's previous behavior. Still, most of us will judge it in nonmoral terms, except that we don't think domestic animals should kill humans (in medieval times, animals were put on trial for acts that went against God's directive of "human dominion"). Let's push the distinction a little further, therefore, by assuming that the camel attacked not a boy but a dog. Such an incident is even less likely to arouse moral emotions. Why? Aren't we perfectly impartial?

The problem is that we are *too* impartial. We're so impartial, in fact, that we don't deeply care about the incident. We may be appalled and feel for the dog, but the incident doesn't activate moral approval or disapproval anymore than a rock hitting another rock. In contrast, as soon as we see two people interact, even people we don't know, we cannot help comparing their behavior with how we think humans *ought* to treat each other. If one slaps the other in the face, we immediately have an opinion: was it deserved, excessive, or cruel? Partly, this is because we ascribe intentions more easily to humans than to animals, but the main reason is that a human scene automatically activates community concern. Is this the sort of behavior, we ask ourselves, that we'd like to see around us, such as helpfulness and mutual support? Or does it undermine the common good, as does lying, stealing, or brutality? We are very conscious of the consequences, and have trouble staying neutral. None of these concerns are triggered by what happens between a camel and a dog.

Chris Boehm, an American anthropologist who has worked with both humans and apes, has insightfully written about the way hunter-gatherer communities enforce the rules. He believes it may lead to active genetic selection similar to that of a breeder who picks animals on appearance and temperament. Some animals are allowed to reproduce, others aren't. Not that hunter-gatherers explicitly think about human genetics, but by ostracizing or killing persons who violate too many rules, or breach one that's too important, they do remove genes from the gene pool. Boehm describes how criminal bullies or dangerous deviants may be eliminated by a member of the community, who has been delegated by the rest to shoot an arrow through their heart. Applied systematically over millions of years, such morally justified executions must have reduced the number of hotheads, psychopaths, cheats, and rapists, along with the genes responsible for their behavior. There are still plenty of such people left, one might object, but this doesn't deny the possibility that there has been selection against them.[28]

It is a fascinating thought that humanity may have taken moral evolution in its own hands, with the result that ever more members of our species are prepared to submit to the rules.

Prozac in the Water

Hunter-gatherer societies don't even allow a hunter to mention what he has killed. According to Richard Lee, the !Kung San hunter arrives in camp without a word, sits down at the fire, and waits for someone to come up and ask what he has seen that day. He then calmly replies something like this: "Ah, I'm no good for hunting. I saw nothing at all (pause) . . . just a little tiny one." Such words make the listener smile to himself, though, for they signify that the speaker must have caught something big.[29] Hunter-gatherer cultures revolve around community and sharing, and stress humility and equality. They frown on anyone with a big mouth. Western society, in contrast, celebrates individual achievement and permits successful individuals to hold on to their gains. In such an environment, humility can be hazardous.

Lamaleran whale hunters, in Indonesia, roam the open ocean in large canoes, from which a dozen men capture whales almost barehanded. The hunters row toward the whale, the harpoonist jumps onto its back to thrust its weapon into it, after which the men stay nearby until the leviathan dies of blood loss. With entire families tied together around a life-threatening activity, their men being literally in the same boat, distribution of the food bonanza is very much on their mind. Not surprisingly, the Lamalera are more sensitive to fairness than most cultures tested by anthropologists, who have played the Ultimatum Game all over the world. This game measures preferences for equitable offers. The Lamalera are the champions of fairness, in contrast to societies with greater self-sufficiency, such as horticulturalists in which every family tends its own plot of land.[30]

If there is such a thing as the moral law, therefore, it is unlikely to be identical everywhere. It can't be the same for the !Kung San, the Lamalera, or a modern Western nation. Our species does possess invariant characteristics, and all of human morality is preoccupied with the two *H*'s of helping and (not) hurting; hence some degree of universality is to be expected. Yet, the details of how fairly resources are divided or how much humility is desirable cannot be captured in a single law. Morality also changes over time within every society, so the hot issues of today may have meant little to previous generations. Sexual mores offer a good example. The Celtic tribes that the Romans encountered while invading northern Europe are said to have been ruled by free-loving queens with scandalous sexual attitudes, at least in the eyes of the patriarchies that followed. Even though this remains hard to substantiate, there is no doubt that the products of these patriarchies were in for a shock centuries later, when Captain Cook landed on the coast of Hawaii. Knowing few sexual constraints, the islanders were characterized as "licentious" and "promiscuous." This scornful terminology is questionable, however, given that there are no signs that anyone got hurt, which for me would be the only reason to reject a given lifestyle. In those days, Hawaiian children were trained through massage and oral stimulation to enjoy their genitals. According to the sexologist Milton Diamond of the University of Hawaii, "The concepts of premarital and extramarital sexual activities were absent, and as in much of Polynesia, no people in the world indulged themselves more in their sensual appetites than these."[31]

Female sexual autonomy is considerably greater in matriarchal than in patriarchal societies, and humanity has experimented with myriad reproductive arrangements. We may have adopted strict monogamy only after the agricultural revolution, about ten thousand years ago, when men began to worry about passing on their daughters and their wealth. Reproductive obsessions with fidelity and virginity may have arisen only at that time. At least, this is the suggestion in *Sex at Dawn*, by Christopher Ryan and Cacilda Jethá, who provocatively take the bonobo as ancestral model of human sex life. In a chapter entitled "Who's Your Daddies?" they explain how in certain cultures a child benefits from having multiple fathers. Their argument rests on Sarah Hrdy's pioneering work on the survival value of multiparental families, including her rejection of the dogma that men will care only for children certain to be their own. Some tribes practice "partible paternity" in which the growing fetus is supposedly nourished by semen of all the men a woman sleeps with. Every potential father claims a piece and is expected to help support the child. This arrangement, which is common in tribes in lowland South America and which guarantees support in an environment with high male mortality, implies reduced sexual exclusivity. A woman's sexual choices outside the marriage are respected rather than punished. On the day of marriage, bride and groom are told to take care of their children, but also to rein in their jealousy of each other's lovers.[32]

Sexual jealousy may well be universal, but its encouragement or discouragement is entirely up to society. So much for universal moral norms. Rather than reflecting an immutable human nature, morals are closely tied to the way we organize ourselves. Nomadic cattle herders cannot be expected to have the same morality as large-game hunters, who cannot be expected to have the same morality as industrialized nations. We can formulate all the moral laws we want; they will never apply everywhere to the same degree. Whether the Ten Commandments are an exception, as is often assumed, is doubtful. Do these commandments even help much with moral decision making? When a conservative politician on a comedy show, *The Colbert Report*, claimed that the Ten Commandments should remain on public display since "without them we may lose a sense of our direction," the host simply asked him to cite them. The politician was taken aback. To the hilarity of the audience, he was unable to comply, except for saying, "Don't lie, don't steal."[33]

Most of the commandments have nothing to do with morality, though, as was pointed out by Christopher Hitchens. They are about respect. In the first five commandments, God insists on exclusive loyalty ("Thou shalt have no other gods before me") as well as respect for our elders. Only after this, does he move on to the "thou shalt not" commands everyone knows. In Hitchens's words:

> *It would be harder to find an easier proof that religion is man-made. There is, first, the monarchial growling about respect and fear, accompanied by a stern reminder of omnipotence and limitless revenge, of the sort with which the Babylonian or Assyrian emperor might have ordered the scribes to begin a proclamation. There is then a sharp reminder to keep working and only to relax when the absolutist says so. A few crisp legalistic reminders follow. . . . But . . . it is surely insulting to the people of Moses to imagine that they had come this far under the impression that murder, adultery, theft, and perjury were permissible.*[34]

The sixth commandment ("Thou shalt not kill") sounds straightforward enough, but if a foreign army were to invade my country or if someone were to abduct my child, I'd have plenty of justification for ignoring this command. The Bible itself lists many exceptions. Capital punishment by legitimate authorities, for example, doesn't seem to fall under it. Clearly, the Ten Commandments were not intended to be taken literally.

The two most popular secular moral laws don't fare much better. Much as I like the sound and spirit of the golden rule—"Do unto others as you would have them do unto you"—it has a fatal flaw. It assumes that all people are alike. To give a rather crude example, if at a conference I follow an attractive woman whom I barely know to her hotel room and jump into her bed uninvited, I can pretty well guess how she'd react. If I explain that I am just doing to her what I would love her to do to me, I'm afraid my appeal to the golden rule won't fly. Or, let's assume that I knowingly serve pork sausages to a vegan. Liking meat myself, I am just following the golden rule, but the vegan will consider my behavior obnoxious, perhaps even immoral. Churchland offers another example, this one of well-meaning Canadian bureaucrats, who removed children from native Indian families to let them be raised by whites. They might have wanted this for themselves, had they been living in camps in the bush, but—like the one that led to the Australian "lost generation" of indigenous children—forced integration policies are now considered grossly immoral. The golden rule doesn't help resolve most dilemmas, such as whether the death penalty is moral or immoral, or whether Jean Valjean, in *Les Misérables*, was right to steal food for his starving niece, or not. The golden rule has a very limited reach, and it works only if all people are of the same age, sex, and health status with identical preferences and aversions. Since we don't live in such a world, the rule really isn't as useful as it sounds.

The second popular secular rule is the greatest happiness principle, also known as utilitarianism, which was recently chosen by Sam Harris as the "scientific" bedrock of morality.[35] Philosophers tumbled over each other to point out that there is nothing scientific about a proposal that came from Jeremy Bentham and John Stuart Mill, two nineteenth-century British philosophers, and goes all the way back to Aristotle. The idea that morality ought to increase "human flourishing" (from the Greek *eudaimonia*), and that good moral decisions will make the maximum number of people happy, is not based on any empirical evidence: it is a value judgment. Value judgments are always up for debate, and the flaws of utilitarianism have been known for a long time. Even though the desire to increase the sum total of happiness in the world will generally nudge us in the right direction, it is far from foolproof. Let's say, I live in an apartment building in which a single man makes over one hundred people miserable by playing the tuba all night, every night. Having failed to dissuade him from producing his din, one of us simply shoots him in his sleep. He never knew what happened. Given how much collective suffering was relieved, what could be wrong with our decision? And if you don't like the shooting part, let's say we gave him a lethal injection. Yes, it deprived one man of his life and possible happiness, but the total amount of well-being in the building clearly went up a notch. In utilitarian terms, we did the right thing.

Other problems with this approach have been pointed out, such as the solution to put Prozac in the water. What a great way to produce a society of happy fools! Or we could follow the North Korean example, and manipulate the media to make everyone feel good about how things are going in the nation, creating a Brave New World of ignorant bliss.[36] All of this would raise the happiness barometer, yet doesn't sound particularly moral. But my own problem with the utilitarian premise goes deeper, and is more serious, because I feel that it runs totally counter to basic biology. I cannot imagine any society, human or animal, without loyalties. All of nature is built around the distinction between in-group and out-group, kin and nonkin, friend and foe. Even plants recognize genetic kinship, growing a more competitive root system if potted together with a stranger rather than a sibling.[37] There is absolutely no precedent in nature of individuals that indiscriminatingly strive for overall well-being. The utilitarian proposal ignores millions of years of family bonding and group loyalty.

One might argue that we would be better off without these allegiances, that we shouldn't worry about who benefits from our behavior and who doesn't. We should simply override our biology in the service of a more perfect general morality. This may sound great, until we consider the flip side of this coin, which is the loss of any kind of commitment and group solidarity. "Family comes first" is not a utilitarian slogan. On the contrary, utilitarianism asks us to subordinate our family to the greater good. I find this impossible to swallow. If all the children in the world have exactly the same value for everyone, who is going to stay up all night next to a sick one, or worry about their doing their homework? Had Valjean been utilitarian, he would have had no pressing reason to bring his loaf of bread home. He might just as well have handed it out to hungry children in the street. The utilitarian position raises shocking questions, such as why I should stay married if another woman needs me more, or why I should help my parents if other seniors are worse off? There would also be nothing wrong with my selling my nation's military secrets, especially not if to a populous nation. If the number of people I'd make happy in the other country would exceed the number I'd make unhappy in my own, I'd be doing the right thing. If my own country doesn't share this opinion, is this just because it is excessively sensitive? I personally don't think so, since for me loyalties are not just morally inconvenient, as utilitarians might call them, but very much part of the moral fabric. We expect them, and are appalled by their absence, such as parental neglect, refusal to pay child support, or treason. We despise the last so deeply that our answer is the firing squad.

I once publicly debated these issues with Peter Singer, the philosopher who is so utilitarian that he feels not even our own species deserves special loyalty.[38] The suffering and happiness of humans and animals are entered into a single equation that covers varying degrees of sentience, dignity, and capacity for pain. The math is mind-boggling. Is one person equivalent to a thousand mice? Is an ape worth more than a Down syndrome human baby? Does a patient with severe dementia have any value at all? After much back and forth, Singer and I found common ground, namely, that humans ought to treat other animals as well as they can. A message of compassion is far more appealing to me than any cold calculation. Singer was forced to concede the drawbacks of his own approach when it came out in the media that he was paying private aides to take care of his mother with advanced Alzheimer's. Asked why he didn't channel his money to more deserving people, at least according to his own theories, he reacted, "Perhaps it is more difficult than I thought before, because it is different when it's your mother."[39] The world's best-known utilitarian thus let personal loyalty trump aggregate well-being, which in my book was the right thing to do.

This brief excursion into the Ten Commandments, the golden rule, and the greatest happiness principle, shows my skepticism that moral dos and don'ts can be captured in simple unassailable rules. Attempts to do so follow the same top-down logic of religious morality that we are trying to leave behind. It is also not free of danger, since it risks leading us down the wrong path, putting principles before people. In an extreme reaction, the normative quest has been labeled "morally irresponsible."[40] Reading Kitcher, Churchland, and other philosophers, one can see an alternative movement underway that tries to ground morality in biology without denying that its specifics are decided by people.[41] This is also my view. I don't believe that watching chimpanzees or bonobos can tell us what is right or wrong, nor do I think that science can do so, but surely knowledge of the natural world helps us understand how and why we came to care about each other and seek moral outcomes. We do so because survival depends on good relations as well as a cooperative society.

Moral laws are mere approximations, perhaps metaphors, of how we should behave. That the underlying values can be internalized to the point that we end up with an autonomous conscience, is something that, as Kant observed, should fill us with wonder, because how this happens is barely understood.

Toeing the Line

Let me close with two more stories about one-on-one morality and community concern in our close relatives. I am not claiming that apes are moral in the sense that we are, but they do show both of these critical ingredients. The first story echoes my earlier remark that bonobos remember where exactly they bit someone and that they show concern, perhaps even regret, about what they did. Instead of a situation among the apes themselves, however, here we consider their reaction to a veterinarian at the Milwaukee County Zoo. While still living in Wisconsin, decades ago, I visited these bonobos many times. They showed all sorts of remarkable expressions of empathy, especially their alpha male, Lody. He was very protective, for example, of an aging female, Kitty. Being blind as well as deaf, Kitty risked getting lost in a building full of doors and tunnels. In the morning, Lody would gently bring her to her favorite sunny spot outside in the grass and by the end of the day wake her up to guide her back indoors, taking her by the hand. If Kitty had one of her epileptic attacks, Lody would refuse to leave her side.[42]

One time, however, Lody was less than empathic, biting the finger of the veterinarian while she was handing out vitamins through the mesh. She tried to pull her hand away, but he bit down. Hearing a crunching sound Lody looked up, seemingly surprised, and released the hand minus a digit. In the hospital, doctors were unable to reattach it. Within days, however, the victim visited the zoo again and, noticing Lody, held up her bandaged left hand, and said, "Lody, my man, do you know what you did?" Lody took one glance at the hand and went to the farthest corner of the exhibit, where he sat head down with his arms wrapped around himself.

In the ensuing years, the veterinarian moved away and rarely stopped by, but fifteen years after the incident, she made an impromptu visit. While she stood among the public, Lody could easily have ignored her, but immediately came running over. He tried to look at her left hand, which was hidden below the railing where he couldn't see it. He kept looking left, insisting to see the hand he had bitten, until the veterinarian held it up. He looked directly at the incomplete hand, her face, and at the hand again. "He knew," she concluded, suggesting that bonobos are quite aware of the consequences of their behavior. This is also my impression, although it is hard to verify, since no one sets up experiments with a fifteen-year time lag. If true, it confirms how deeply these apes care about their relationships, showing the sort of concern that underlies human moral tendencies.

The second story took place at the Arnhem Zoo while I was still there. One balmy evening, we called the chimps inside. But because the weather was superb two adolescent females refused to come in. They enjoyed having the island to themselves. The rule at the zoo being that none of the apes would get dinner until all of them were inside, the obstinate females caused a grumpy mood. When they finally did enter, hours late, they were assigned a separate bedroom so as to prevent reprisals. By the next morning, when all of us had forgotten about the incident, the chimps showed they had not. Once out on the island, the whole colony vented its frustration about the delayed meal by a mass pursuit ending in a physical beating of the two culprits. Admittedly, the rule they had violated was human imposed, but perhaps this is what helped us appreciate its enforcement. The colony seemed to grasp the benefits of having everyone toe the line.

That evening, the two teenagers were the first to come in.

Chapter 7

THE GOD GAP

If God didn't exist, he'd need to be invented.

—Voltaire[1]

Ironically, one of the least empathetic men in history, Mobutu Sese Seko, preserved the jungle that now serves as the playground of the world's only bonobo sanctuary. It was the last piece of jungle in the capital city of Kinshasa. At Lola Ya Bonobo (Lingala for "bonobo paradise"), a large number of apes live on the lush grounds of the former Congolese dictator's weekend retreat. It is here that the man with the leopard skin hat, who extracted anywhere between five and ten billion dollars from this poor nation, feasted on delicacies flown in from Europe while plotting the public hangings of his rivals.

The Democratic Republic of the Congo (DRC) is a huge country—the size of Western Europe—that encompasses the bonobo's native habitat. The species is gravely endangered, however, with only an estimated 5,000 to 50,000 bonobos left. Even the latter figure is still less than the number of seats in a typical sports stadium. Unfortunately, wild bonobos are being killed for bush meat, while any babies found clinging to the victims are kept alive, since they fetch thousands of dollars on the black market. The selling of bonobos is illegal, though, so these baby apes are often confiscated and brought to Claudine André, the Belgian founder and director of the sanctuary. At Lola, the orphans are raised in a nursery by *Mamans*, local women who watch over them and give them the bottle. After a few years, the youngsters join the groups out in the forest, which roam on their own even though they remain dependent on human provisioning. It is here that we conduct our studies of empathy, since we can get much closer to these apes than to those in the wild. Fieldworkers are lucky to see their bonobos regularly, and sustained observation of social interactions is nearly impossible in dense foliage.

One of my co-workers, Zanna Clay, patiently waits for spontaneous conflicts among the bonobos, collecting them on video so that we can analyze their aftermath. Inevitably, these incidents cause distress in one or both parties. How do bystanders react? They reassure anyone who has lost a fight by means of genito-genital rubbing, a brief mount, or a manual massage of the genitals. What chimps do with platonic touching calls for sexual engagement among bonobos. The principle, however, is exactly the same in both species: the apes down-regulate each other's anxieties. This is such a basic emotional response that we notice it even among orphans in the nursery, who have barely had any social models to learn it from. And they, too, often do so in a sexual manner.

At other times, however, bonobos behave more like chimps, embracing or grooming each other. Take the return of Makali, an adult male, who was injured in a fierce group attack. His hand was badly bitten. Afterwards, Makali hid from the group for several nights, biding his time in the forest instead of joining the rest in their sleeping quarters. When he finally did show up, he quietly sneaked into a group lounging in a shady area. He was immediately surrounded by curious juveniles. Some seemed to mimic the awkward way in which he held up the limb with its infected wound. They reached for it, but Makali systematically avoided their grabby little hands. With a pained expression on his face, he held his hurt finger in front of him, bending his wrist downward. The adults were more diplomatic, initiating contact with grooming. He was first approached by a dominant male, who planted a kiss on his neck, after which he was groomed by one of the females. Whereas the juveniles had come up to him in gangs, the adults seemed to await their turn, arriving one by one. The first to handle his injury was the alpha female, Maya. She groomed him briefly, then took his hand and carefully licked the open wound. He let her do so. This seemed to mark his acceptance back into the group, leading one of his main attackers to enter the scene. Normally, Makali had an entirely antagonistic relationship with this male. But after briefly staring at the injury, he groomed Makali, something no one remembered ever having seen before.[2]

These are the normal love-hate cycles of social life. The same alternation between conflict and reunion marks human families, marriages, and every typical primate group. It is good to keep in mind, though, that these bonobos are far from typical. They have suffered unimaginable abuse at the hands of humans, losing their mothers to poacher snares or bullets at a tender age. That they are capable of reconciling after fights at all, and of calming down upset companions, is truly remarkable. Zanna has noticed that bonobos born in the Lola groups (the apes are allowed to breed, and some orphans have become mothers) are far more skilled at conflict resolution and more inclined to show sympathy than their orphaned peers. This advantage of mother-rearing fits what we know about emotion regulation, also in humans. Romanian orphans, for example, are marked by long-lasting emotional devastation. This makes it all the more astonishing that the orphans at Lola have built a decent social life together. It is a testimony of their resilience and of the great human care they have received. After having lost everything at the hands of hunters, they were lovingly fed and cuddled by humans, who became substitute mothers. They had to mentally distinguish between the contrasting manifestations of this bipedal ape, capable of such cruelty and such kindness at the same time. It seems a confusing lesson to learn early in life.

Having grown up in a nursery, the bonobos remain fascinated by bottles and employ them to demonstrate their empathy. One adult female picks up an empty plastic bottle, fills it up with muddy water from the river, and sits down in front of two juveniles, one of whom is her own child. She then moves the bottle gently to the mouth of one and tips it so that the water flows between his pouted lips, filling up his lower lip. Once his mouth is full, the female tips the bottle slightly back and keeps it in place while waiting until he has gulped. Then she gives him some more, until she turns her attention to the other juvenile, who pouts as soon as she looks at her, knowing it is her turn. The female puts the bottle near her mouth, too, and starts over. I have never seen this procedure, full of gentle attention to the other's swallowing abilities, from any other ape. Possibly, the female is reenacting the role of *Maman*, and the juveniles play along.

Since the whole river runs right next to them, it can't be about needing a drink.

Life and Death

Awareness of death is often mentioned as a reason why we humans developed religion. Our sense of mortality is mentioned in one breath with the question whether we may be the only ones to have it. I have no clear answer, except that when it comes to the mortality of *others*, there is no reason to assume ignorance in our fellow primates. Like the Lola bonobos, apes are plenty familiar with death and loss. Sometimes they are themselves the killers, as on the day the bonobos dispatched a Gaboon viper, a highly poisonous snake. The snake aroused fear; everyone jumped back at its every move. They carefully poked at it with a stick, until Maya threw it in the air and slammed it against the ground. Strikingly, after the snake had been killed, no one gave any indication of expecting it to come back to life. Dead is dead. The juveniles happily dragged its lifeless corpse around as a toy, slinging it around their necks, even prying open its mouth to study its huge venomous fangs.

The scene reminded me of a chimpanzee hunt I once observed. Following the apes around in the Mahale Mountains, in Tanzania, we heard a sudden commotion high up in the trees. Chimpanzees have a special scream when they have captured prey. The existence alone of such a vocalization indicates their willingness to share meat, because otherwise they'd be wiser to stay silent. The screams attracted many others. Several males had caught a red colobus monkey, the sort of prey chimps have trouble capturing on their own. It is usually a team effort. Staring up through layers of branches and leaves, I saw that the hunters were starting to eat while the monkey was still alive. Since chimps are no "professional" predators, they never evolved the effective killing techniques of the cat family, and—as is also true of humans—their treatment of prey reflects poorly on their empathic capacities. Many chimps gathered in a feeding cluster, including females with swollen genitals, who tend to enjoy priority. It was all very noisy and chaotic, but everyone ended up with a piece of monkey meat. The next day, I noticed a female chimp walking by with a juvenile jockey riding on her back. Her daughter happily swung something fluffy in the air, which turned out to have belonged to the poor monkey. One primate's tail is another's plaything.

One morning, Geza Teleki followed a party of chimps, hearing raucous vocalizations in the distance. Six males were wildly charging about, uttering "wraaah" calls that echoed off the valley walls. In a small gully lay the motionless body of Rix sprawled among the stones. Although Teleki had not seen the body drop, he felt he was witnessing the initial reaction to this male's neck-breaking fall out of a tree. Several

Dorothy, a thirty-year-old female chimpanzee, died of heart failure at a sanctuary in Cameroon. Staff brought her out in a wheelbarrow to display her body. The normally rowdy chimps gathered around, staring at the corpse and holding on to each other. They were as silent as people at a funeral.

individuals paused to stare at Rix's corpse, after which they vigorously charged away from it, hurling big rocks in all directions. Amid the noise, chimps were embracing, mounting, touching, and patting one another with big, nervous grins on their faces. Later on, the chimps spent considerable time staring at the body. One male leaned down from a limb, watched the body, and whimpered. Others touched or sniffed Rix's remains. An adolescent female gazed at his body without interruption for over one hour, motionless and in silence. After three hours of activity, one of the older males finally left the clearing, walking downstream. Others followed one by one, glancing over their shoulder toward the corpse as they departed.[3]

Reports of how apes respond to death are becoming more numerous. In 2009, a photograph went viral of the death of Dorothy, whose corpse aroused intense (but eerily silent) attention from her sanctuary community. In the Blair Drummond Safari Park, in Scotland, the death of an elderly female, named Pansy, was carefully analyzed from video. In the ten minutes before her death others groomed or caressed

Pansy a dozen times, and Pansy's adult daughter remained with her throughout the night. Reactions to her death ranged from testing her mouth and limbs, perhaps to see whether she was still breathing or able to move, to a male who slammed her body. This behavior has also been seen after other deaths. It comes across as insensitive, but may be a way of trying to rouse the dead. Apes often react with a combination of frustration about the lack of response and testing whether a response can be provoked. Most of the gathering individuals are subdued, however, as if they realize that something terrible has happened. From their observations of Pansy's final hours, the investigators concluded that "chimpanzees' awareness of death has been underestimated."[4]

An adolescent female, Oortje, literally dropped dead at the Arnhem Zoo. I knew Oortje as a happy character, playful and gentle with flapping ears (her name means "little ear"). She had been very quiet the last few weeks, however, and had started coughing. Her condition deteriorated despite antibiotics. Kept in its winter quarters, the colony was divided into two groups, which could hear but not see each other. In the middle of the day, one adult female was seen staring into Oortje's eyes from up close. Without any apparent reason, this female burst out screaming in a hysterical voice while hitting herself with spasmodic arm movements, as frustrated chimps often do. The female seemed profoundly upset about something she had detected in Oortje's eyes. Oortje herself had been silent until this point, but now feebly screamed back, then tried to lie down, fell off the log on which she had been sitting, and remained motionless on the floor. A female in the other hall uttered screams similar in sound to those of the first, even though she could not possibly have seen what happened. After this, twenty-five chimps in the building turned completely silent. Autopsy showed that Oortje had suffered a massive infection of her heart and abdomen.

In general, the reactions of apes to the death of companions suggest that they have trouble letting go (mothers may carry dead infants around for weeks, until the corpse is dried out and mummified), test the corpse, try to reanimate it, and are both upset and subdued. They seem to realize that the transition from alive to dead is irreversible. Some of the reactions resemble the way humans attend to their dead, such as the touching, washing, anointing, and grooming of bodies before we put them into the ground. Humans go further, though, in that they often give the dead something on their "voyage," such as the Egyptian pharaohs whose tombs were filled with ample amounts of food, wine and beer, hunting dogs, cats, pet baboons, and even full-sized sailing vessels. Humans often look at death as a continuation of life. There is no indication that any other animal does so.

Apes do seem to worry about the possible death of others. If bonobos get caught in poacher snares intended for bush pigs and duikers, they usually manage to free themselves, but there are enough wild bonobos with missing fingers or hands for us to conclude that they aren't always so lucky. Upon hearing sudden screams in the swamp forest, fieldworkers found a male, Malusu, crouching with a metal snare around his hand, dragging a sapling to which the snare was attached. The sapling impeded his locomotion. Other bonobos unfastened the snare from the lianas, and tried to remove it from Malusu's hand. He kept getting stuck, however, and was left behind while the others traveled to the dry forest where they usually slept. The next morning, these bonobos did something never observed before: they returned over a one-mile distance straight to the same spot where they had last seen Malusu. Once there, they slowed down and searched around. Given their knowledge of snares, the bonobos may have made the connection with the loss of a group member. They failed to find Malusu, but a month later he rejoined the community. Despite a permanently mangled hand, he had survived his ordeal.[5]

It seems safe to say that apes know about death, such as that it is different from life and permanent. The same may apply to a few other animals, such as elephants, which pick up ivory or bones of a dead herd member, holding the pieces in their trunks and passing them around. Some pachyderms return for years to the spot where a relative died, only to touch and inspect the relics. Do they miss the other? Do they recall how he or she was during life? While such questions are impossible to answer, we are not the only ones fascinated and intimidated by death.

We once tested the permanency of death on the Arnhem chimpanzees by showing them their former friends and rivals. A wonderful film, *The Family of Chimps*, had captured the apes' personalities and intelligence as no documentary had ever done before. It was a television hit all over the world.[6] I had left the Netherlands before the movie was made, and watched it the first time with tears in my eyes because of the loving attention with which all of my old friends had been portrayed. The film showed Nikkie as the alpha male of the colony, but in the years that followed two males moved closer to a coalition against him. Nikkie must have been on edge as never before, because one morning, hearing screaming and hooting behind him, he raced full speed out of the building, straight toward the moat surrounding the island. A year earlier, Nikkie had managed to cross this moat thanks to a layer of ice. Perhaps he thought that he could repeat this feat. This time, however, he failed and drowned. The newspapers dubbed it a "suicide," but it more likely was a panic attack with fatal outcome.

With Nikkie's death, the closeness between the two other males evaporated. Rivalry took its predictable place. Dandy became the new alpha, but the ghost of Nikkie still lingered, as revealed by the colony's reaction to *The Family of Chimps*. One evening, more than two years after its making, the winter hall was turned into a theater. With all the lights dimmed, the movie was projected onto a blank wall. The apes watched in complete silence, some with their hair fully on end. When, in the movie, a female was attacked by pubertal males, several indignant barks were heard, but it remained unclear whether the apes recognized the actors. Until, that is, Nikkie appeared in full glory on the wall. Dandy bared his teeth and ran screaming to the male who had supported him against Nikkie, literally jumping into his lap. The two males embraced each other with big, nervous grins on their faces.

Nikkie's "resurrection" had revived their old partnership.

Dancing in the Rain

Proposed origins of religion are a dime a dozen. Fear of mortality is just one of them; there are plenty more. According to one theory, which sounds as if it was invented in a bar, intoxication is at its root. Wine and beer have traditionally been thought to fortify the body, but they also feed the imagination. In an act of self-aggrandizement typical of drunks, our ancestors began to picture themselves as invincible and look beyond their immediate existence. This mind-altering connection is still visible in the role of "spirits" (the term alone!) in religious rituals, such as those of the Greek Dionysus wine cult, Catholic mass, which features wine as Christ's blood, and the Kiddush, a Jewish blessing recited before drinking: "Blessed are You our God, creator of the fruit of the vine." Mentioned 231 times in the King James Bible, wine is central to many religions for its miraculous quality of releasing the human spirit.

The health benefits of fermented beverages, and concern about our bodily condition in general, were very much part of early religions. Without recourse to effective medicine, everyone could die of a minor infection. People turned to religion to find solace and pray for cures. They may have been right to do so, given the well-established epidemiological connection between religiosity and health.[7] Religion seems to promote well-being in body and mind. Let me hasten to add, though, that there is little agreement about how it does so. Even if many religions have rules governing diet, drugs, marriage, and hygiene, this doesn't seem the reason. Research points, instead, to church attendance as a major factor, which suggests a social dimension. It is well known that social connectedness strengthens the immune system, and church attendance surely helps in this regard. If so, it may not be religiosity per se that protects against disease, but rather human contact. For all we know, the same benefits may apply to members of a book club or birding society. Churches, however, produce more shared commitment, which does add to a sense of belonging. Émile Durkheim, the French father of sociology, emphasized the collective rituals, sacred music, and singing in unison that make religious practice an irresistible bonding experience. Others have depicted God as an attachment figure, who offers safety and comfort in stressful situations. In addition, many religions add female statues marked by a gentle, nonjudgmental facial expression. These maternal sources of solace—from Mary in Christianity to Demeter in Greece and Guanyin in China—are designed to lighten our load of sorrow the way mothers do for their children.

But origin stories of religion don't end here. There is also the awe and wonderment at natural events beyond our control. That this may not be uniquely human is illustrated by the charging displays of chimpanzees at waterfalls or during downpours. The first time I witnessed this, I had trouble believing what I saw. The chimps at the Arnhem Zoo sat around miserably with their "rain faces" (an expression of disgust with eyebrows pulled down and lower lip stuck out) under the tallest trees, doing their best to stay dry. When the rain intensified, however, and reached under the trees, two adult males got up, with bristling hair, and started a display known as the bipedal swagger (which, one can imagine, made them look human in a thuggish sort of way). With big, rhythmic, swaying steps, they walked around, leaving their shelter, getting completely wet. They sat down again when the rain eased. Having seen the same behavior several times since, I agree with those who call it a "rain dance," because that's exactly what it looks like. Jane Goodall described a chimpanzee male acting similarly near a roaring waterfall:

As he gets closer, and the roar of the falling water gets louder, his pace quickens, his hair becomes fully erect, and upon reaching the stream he may perform a magnificent display close to the foot of the falls. Standing upright, he sways rhythmically from foot to foot, stamping in the shallow, rushing water, picking up and hurling great rocks. Sometimes he climbs up the slender vines that hang down from the trees high above and swings out into the spray of the falling water. This "waterfall dance" may last ten or fifteen minutes.[8]

Goodall went on to wonder whether these displays could become ritualized into some animistic religion, and what would happen if chimps could share these feelings with each other. Would it lead to collective worship of the elements? An entirely different take on the same behavior, however, would be that the apes believe, for whatever reason, that they can influence the course of nature. Perhaps a fortuitous event, such as the ceasing of rain in the middle of a charging display, created the superstitious belief that if they display hard enough, they can stop precipitation. For those who regard such mistaken associations as dim-witted, it is good to realize that there is little doubt which ape is the most superstitious of all, and it's not the chimpanzee.

Young chimps are smarter than children. At least, this was the shocking conclusion from an experiment in which scientists demonstrated a simple procedure to both chimps and children. A scientist poked a stick into holes in a large plastic box, going through a series of holes until candies rolled out. Only one hole mattered, however—the other holes had nothing to offer. If the box was made of black plastic, it was impossible to tell that some of the poking was just for show. With a transparent box, on the other hand, it was obvious where the candies came from. Handed the stick and the box, young chimps mimicked the necessary moves, while ignoring all empty holes, at least if the box was transparent. They had paid close attention. The children, however, mimicked everything the scientist had shown, including moves without any purpose. They did so even with the transparent box, approaching the problem more like a magical ritual than as the goal-directed task that the apes had seen in it.[9]

Our species is incredibly superstitious; it develops lots of habits unworthy of a rational animal. We knock on wood when we don't want to tempt fate, we wear a worn-out T-shirt for luck during our team's matches, some soccer players won't enter the field without wearing their underwear inside out, and baseball players go through a dozen rituals before they even pick up a bat. Turk Wendell always chewed four pieces of black licorice while pitching. He'd spit them out at the end of every inning, after which he would return only after having brushed his teeth. We're also sensitive to numbers, with the Chinese systematically avoiding 4 and Westerners fearing 13. When I came to the United States, I was struck by the omission of the thirteenth floor in buildings, but am now so used to it that I was shocked in the reverse cultural direction when recently, on a large Dutch cruise ship, they recommended lifeboat 13 for my safety. The composer Arnold Schönberg suffered from triskaidekaphobia to such a degree that it may have spelled his death. He was especially afraid of multiples of 13, and then, right before his sixty-seventh birthday, a friend pointed out that numbers could also be added up. What are friends for? Staying in bed the whole day, Schönberg almost got through his birthday, but his heart stopped beating a quarter before midnight on 13 July 1951. A Friday, moreover.

Some house cats seem to think that they will get fed if they scratch the couch, and some dogs turn circles in the kitchen because in the past they have received food while doing so. There are also negative associations. One of our cats, Loeke, had surgery near his anus and was in such pain each time he defecated that he started to "blame" the litter box. He'd approach the box only if he couldn't wait any longer, almost stalking it, then racing into and out of it with a speed as if the thing might attack him. We were very patient, and cleaned up a lot of mess for half a year, to get him over his phobia. Such spurious associations were labeled "superstitions" by B. F. Skinner. In experiments on pigeons, he had employed an apparatus that delivered food pellets at regular intervals without any relation to the birds' behavior. Spontaneously, the pigeons began to connect the appearance of food with actions they had just performed, so that soon some of them turned little rounds and others pushed their head all the time into the same cage corner. Whether this sort of behavior is truly equivalent to human superstition is debatable, however.

We take superstitions so seriously that they sometimes hamper progress. A classical example is the lightning rod of Benjamin Franklin, one of the founding fathers of the United States. He first used a kite to demonstrate that lightning is electricity, then invented a way of conducting its energy into the ground to avoid damage. Since lightning often struck church towers, the ideal place to put his metal rods was on top of them. But Franklin's focus on churches set him up for a collision with the view of lightning as proof of God's displeasure. Mounting his device on the house of God was like defying his will. *Stealing God's Thunder* is the apt title of one book about Franklin.[10] His rods were highly effective, though. Fewer churches were destroyed around Boston, where most of the devices were mounted, than anywhere else. Nevertheless, some found the whole idea of escaping the hand of God sacrilegious. When Massachusetts was in 1755 hit by a major earthquake, a pastor accused Franklin of having invited the quake by his heretical arrogance.

Superstition blurs the line between reality and imagination as does religion and a belief in God. At one level, God's existence is an absolute certainty for many, but at another level it always remains open to criticism. Religion is called "faith" precisely because it trusts things unseen. We humans have a knack for this, as was shown in the above imitation experiment with the boxes. While the apes took the task at face value, ignoring unnecessary moves, children put their trust in the experimenter, copying every action. They invested the procedure with mysterious significance. Naturally, psychologists were unhappy with the implication that this made the apes more rational. They were quick to speak of "over-imitation" by the children, seeing this as a good thing. In fact, it's brilliant! Given the superior knowledge of adults, children should copy them without questions. Blind faith, it was concluded, is the more rational strategy.

This doesn't mean that imagination and make-believe are out of the question for our primate relatives. There are reports of human-reared apes, such as Washoe, who carefully bathed her doll, and another ape, Viki, who pretended dragging an imaginary toy around by an imaginary string that she would unhook if it got "stuck." I have already mentioned the bonobo female who fed juveniles from a bottle even though there was no need for it, perhaps imagining she was a *Maman*. In wild chimpanzees, there are observations of care for imaginary young. Richard Wrangham observed a six-year-old juvenile, Kakama, carry and cradle a small wooden log as if it were a newborn. Kakama did so for hours on end, one time even building a nest in a tree and gently placing the log into it. The fieldworker was reluctant to draw conclusions from what he had seen, but had to admit it was a young male playing with a doll. Kakama may have been anticipating a sibling, because his mother was pregnant at the time. I myself have seen juvenile chimps act the same, tenderly holding a piece of cloth or a broom. A wild gorilla was seen to pull up a mass of soft moss, which she carried and held like an infant under her breast, seemingly "nursing" it.[11]

Perhaps apes, too, can create a new reality that exists alongside the old one. In the old reality, a wooden log is just a log, whereas in the new one, it's a baby. This capacity for dual reality is so highly developed in our own species that a sugar pill improves our health even if the nurse takes it out of a bottle with "placebo" clearly written on it. On one level, we know the pill is fake; on another, we still believe it will work. In the same way, we fall for romances, rivalries, and deaths in movies while at the same time well aware that the actors are just acting. We are great at suspending one reality for a new one. It is part of the success of Hatsune Miku, a popular Japanese pop star who draws masses of rocking youths even though she is a mere hologram. She is a computer-generated 3-D projection with a female persona and a synthesized voice, who dances and sings to a live band, towering high above her audience since she is not limited by human size. Her concerts sell out in a matter of minutes. The public sings along with her and responds to her sexy moves as if she were real.

To insist, as neo-atheists like to do, that all that matters is empirical reality, that facts trump beliefs, is to deny humanity its hopes and dreams. We project our imagination onto everything around us. We do so in the movies, theater, opera, literature, virtual reality, and, yes, religion. Neo-atheists are like people standing outside a movie theater telling us that Leonardo DiCaprio didn't really go down with the *Titanic*. How shocking! Most of us are perfectly comfortable with the duality. Humor relies on it, too, lulling us into one way of looking at a situation only to hit us over the head with another. To enrich reality is one of the most delightful capacities we have, from pretend play in childhood to visions of an afterlife when we grow older.

Some realities exist, others we just like to believe in.

No Thought of the Morrow

Borie, an old chimpanzee with a suspected ear infection, made an odd request. While visiting her in the sleeping quarters, she kept waving her hand in the direction of a table. The table was empty, except for a small plastic baby mirror. After a few minutes of this, we thought that Borie might want the mirror and gave it to her.

She took it in one hand and picked up a straw with the other, angling the mirror such that she could look at her ear while poking the straw into it. While cleaning her ear, she carefully followed progress in the mirror as if this had been her intention all along. However simple the task may seem, it required some brain power. First of all, Borie needed to know that she could see herself in a mirror, which is knowledge few animals possess. Mirror self-recognition is well documented in apes, however. Second, she must have planned her move all along, because otherwise why did she put the mirror to immediate use?

It is often thought that animals are captives of the here and now, but Borie must have been waiting to direct us to what she needed. Planning is in fact well developed in the apes. Other examples include wild chimps collecting a bundle of tall grass stems, which they carry in their mouths for miles until they arrive at the termite hills where the stems serve as fishing tools. Similarly, zoo chimps may gather armfuls of straw from their night cage before going outside where it is cold. But the best-known case for planning is undoubtedly provided by Santino, a male chimp at a Swedish zoo. Every morning, before the visitors arrived, he'd leisurely collect rocks from the moat around his enclosure, stacking them up into neat little piles hidden from view. This way, he'd have an arsenal of weapons when the zoo opened its gates. Like so many male chimps, Santino would several times a day rush around with all his hair on end to intimidate the colony. Throwing stuff around would be part of the show, including projectiles aimed at the public. Whereas most chimps find themselves empty-handed at the critical moment, Santino had prepared his rock piles for this purpose. He had done so at a quiet moment, when he was not yet in the adrenaline-filled mood to produce his usual spectacle.[12]

Experiments on planning go back to Wolfgang Köhler, who in the 1920s presented apes with a banana hanging from the ceiling while providing them with boxes and sticks. As we have seen, elephants are also capable of solving this problem. More recently, apes have been given a choice of tools that they couldn't use right away, but might be able to employ later on. The apes preferred these tools over immediate rewards, patiently holding on to them in anticipation of future payoffs.[13] In one innovative test, scientists wanted to see whether the apes could imagine a solution they'd never seen before. They presented them with a transparent container that was too narrow for the ape to reach into. An unshelled peanut was put at the bottom. Unable to access the peanut, and having no tools, what could the ape do but stare at it? But the ape had a solution. He went to the faucet, collected a mouthful of water and spit it into the container. One mouthful of water wasn't enough, though, so the ape had to walk several times back and forth between the faucet and the container to be able to retrieve the floating peanut with his fingers. One ape was even more creative, achieving the same goal via urination.[14]

Knowledge of the future and awareness of death could potentially combine into a sense of mortality. But even if our primate relatives share some of our imagination and future orientation, it remains unclear whether they ponder their own death. An illustrative case involves Reo, a chimpanzee at the Primate Research Institute of Kyoto University. When Reo was in the prime of his life, he became paralyzed from the neck down as the result of a spinal inflammation. He could eat and drink, but couldn't move his body. His weight continued to drop, and he developed extensive bedsores while veterinarians and students cared for him around the clock for six months. Reo recovered, but the most interesting part is how he reacted to his bedridden predicament.

> *One thing very clear to everyone involved in his care was that Reo's attitude to things did not change throughout his period of complete paralysis. He often teased young students by spitting water at them—just as he had done before his illness. His outlook on life was the same after he fell ill as it was before; we did not notice any perceptible change even as he lay there thin as a rake and covered in sores. Bluntly put, he did not seem worried about his future. He*

did not become depressive, even though the situation looked, to us, extremely grave.[15]

Our vaunted imagination is like a double-edged sword. On the one hand, it causes desperation in a situation in which an ape might remain unworried, but it also gives hope as it allows us to envision a better future. We look so far ahead, in fact, that we realize that our life will come to an end. This realization thoroughly colors our existence, leading to a permanent search for meaning as well as bitter jokes along the lines of "Life's a bitch, and then you die!" Would we have developed a belief in the supernatural without this cloud hanging over us? A partial answer comes from research showing that the more aware people are of their own mortality, and the more they think about it, the more they believe in God.[16] They feel the rocking of the boat, so to speak, and, like most voyagers on stormy seas, appeal to a higher power.

But before we conclude that thanatophobia sets us apart, a major qualification is in order, one that I have to think of each time I see Bosch's *Garden.* Most of the time, instead of considering our death, we put thoughts of it on hold. No sane persons literally deny their mortality, of course, but many of us act as if we'll live forever. Bosch's painting presents one giant warning against this illusion. *The Garden* brims with middle-aged singles preoccupied with their little pleasures. "With no thought of the morrow," as one expert put it, "their only sin the unawareness of sin."[17] Despite participating in a massive gathering, the naked figures seem lonely and inward looking, not unlike contemporary teenagers, who remain glued to their individual smartphones while moving around in bands. The hedonists of *The Garden*'s central panel neither raise children nor produce anything of value, locked up as they are in their existential cocoons except for the occasional erotic fling that requires a partner. If this is a paradise of lust, it is one empty

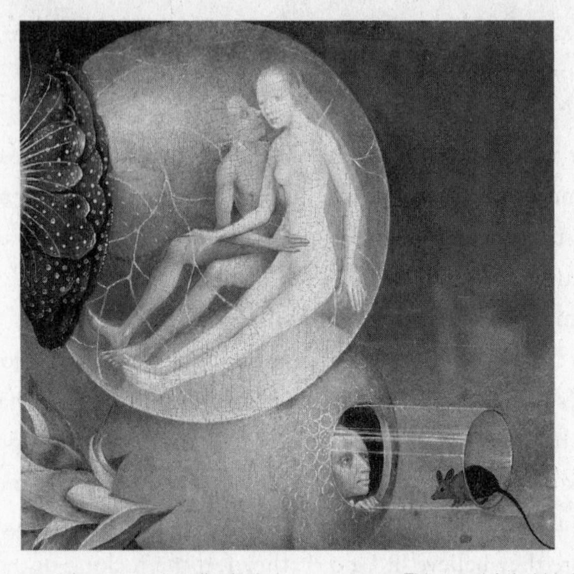

The Garden's many pleasure-seekers live in their own cocoons. This couple's bubble has been interpreted as made of cracked glass to symbolize the fragility of love. But the cracks rather seem to be veins in an amniotic sac, which shields them from the outside world. The man on the right, on the other hand, is looking through a glass tube (an alchemical allusion) at a rat. The symbolism is uncertain, but I can't help seeing a behavioral scientist.

of purpose and accomplishments. They appear oblivious to the larger world, including the death and destruction that will inevitably befall them. They act like immortals. We the viewers, on the other hand, see the horrifying right-hand panel and know what's around the corner.

When a radio show invited the Turkish writer Elif Şafak to give us her "sixty second idea to change the world," she took her inspiration from the Sufi recommendation to taste death before you die.[18] Buddhism knows the same emphasis, seeing it as liberating to accept your own death. Since the modern world is based on a negation of death, Şafak said, we all need to visit a salon, like a hair salon, where for one hour we taste death, our own death. This will produce a milder heart and a greater appreciation of life, she claimed. Although we realize that we're mortal, we have trouble building this knowledge into our lives. Listening to her proposal, I thought it was great for the middle-aged, but unneeded for anyone my age. Members of my generation have either seen our parents die or are prepared for this to happen any moment. We have lost siblings, friends, spouses, perhaps even children. We have friends afflicted with Parkinson's, cancer, Alzheimer's, or some other dreadful disease. The older we get, the more we experience the physical ravages of our own aging, and become acutely aware that our time on earth is limited.

Pieter Brueghel the Elder made this point as explicitly and morbidly as possible in a single painting. We see wagonloads full of skulls, while people of all backgrounds—from farmers to bishops to noblemen—are taken indiscriminately to the other side. The dead advance on the living like an unstoppable army, herding them into huge traps while fires rage in the distance. There is a hanging on the horizon, a dog eats a dead woman's face, and a man with a millstone around his neck is being held above the water. Dinner guests around a table resist in vain, by drawing a sword or running from the advancing corpses, while in the lower right-hand corner an oblivious lover plays the lute to a woman behind whom a skeleton cheerfully plays along. Brueghel's sickening army of death and destruction dates from 1562, half a century after *The Garden*, from which it took its hell-on-earth inspiration. It is appropriately entitled *The Triumph of Death.*

The modern equivalent of this painting is the British artist Damien Hirst's display entitled *The Physical Impossibility of Death in the Mind of Someone Living*. The work presents the body of a tiger shark in formaldehyde in a vitrine; the shark is so large and has such a wide open mouth full of teeth that it brings the prospect of death quite close. When the shark landed in New York, in the Metropolitan Museum of Art, it was characterized as "simultaneously life and death incarnate in a way you don't quite grasp until you see it, suspended and silent, in its tank."[19] But the artwork has also been described as an outrageously expensive fish without chips.

Death is so hard to accept that we try not to think about it, and act as if the deceased will be traveling to a better place, where we will one day meet her or him. Elaborate burials go back to our Cro-Magnon ancestors, who sent off the dead with adornments such as ivory beads, bracelets, and necklaces. No people would leave so many valuables in a grave unless they believed in an afterlife. We are the only species to follow these rituals and enjoy the solace they provide, yet I am still not entirely convinced that we are the only ones with an inkling of our own death. Being a young adult male, Reo may not have been the best example. Impending death is rarely met with acceptance at that age. In many species, aging individuals appear quite a bit wiser than the young, and losing one's vitality slowly over the years may be experienced quite differently from the sudden immobilization that befell Reo. When an old ape notices that trees are harder and harder to get into or an elephant has ever more trouble keeping up with the herd, might these individuals not apply what they have learned about life and death to their own bodies?

It's hard to know, yet impossible to rule out.

Freud's Cold Feet

To delineate religion to everyone's satisfaction is hopeless. I was once part of a forum at the American Academy of Religion, when someone proposed we start off with a definition of religion. However much sense this made, the idea was promptly shot down by another participant, who reminded everyone that last time they tried to define religion half the audience had angrily stomped out of the room. And this in an academy named after the topic! Let us therefore simply say that religion is *the shared reverence for the supernatural, sacred, or spiritual as well as the symbols, rituals, and worship that are associated with it.* This definition lacks a distinction between spirituality and religion, although by insisting on "shared" reverence, it excludes individual approaches and considers only group phenomena. Thus defined, religion is a human universal.

The only exceptions ever mentioned are the Pirahã. But the claim that these Brazilian forest people lack religion (they have been called an "atheist tribe") doesn't survive scrutiny of the original sources. Daniel Everett, the American ex-missionary who lived among the Pirahã, explains how these people talk to and dance for spirits. They wear necklaces made from seeds, teeth, feathers, and beer-can pull tabs, which are "decorative only secondarily, their primary purpose being to ward off the evil spirits that they see almost daily."[20] Not only do they see spirits; they channel them while talking in falsetto voices. The Pirahã are so terrified of evil spirits, though, that they refuse to mention them by name. Even after having just channeled them, they deny their presence ("I don't know, I didn't see it"). Their fear makes it nearly impossible for Westerners to figure out what the Pirahã actually believe in, but there is no doubt that they believe in something. It just isn't what we are used to.

If religion is so widespread, the next question is why it evolved. Biologists always wonder about survival value. What sort of advantage does religion confer? This question has been addressed by comparing early Christians with the Roman population around them. When two plagues swept through the empire, each one killing one-third of the population, the Christians fared better than the Romans. The Christians brought food and water to those too sick to care for themselves, attending to their needs in the name of Christ, whereas the Romans fled from their dearest, abandoning them even before they had expired in an attempt to avert contagion. Even though Christians risked contamination, a study of tombstone inscriptions revealed that they enjoyed a higher life expectancy.

But is this the right comparison? The first flaw is that the Romans were plenty religious themselves, eager to placate and please their gods and goddesses, such as Mars and Venus. We're therefore not really comparing people with and without religion. The second flaw is that the early Christians were not your typical population: they were a persecuted minority, hence part of a close-knit community fighting a common enemy. This must have given them a common purpose, which may have had beneficial health effects. Unfortunately, trying to pinpoint the success of religion is like asking what good it is to have language. I am sure language has its benefits, but since all humans have one we simply lack comparison material. With religion, we're in the same boat. The only thing we do know is that attempts to abolish or discourage religion have had disastrous consequences.

This was true of Stalin in the Soviet Union, Mao Zedong in Communist China, and Pol Pot of the Cambodian Khmer Rouge, all of whom tortured, killed, and starved millions of their own people. The Khmer Rouge banned all religion, while offering the following chilling slogan to the condemned masses: "To keep you is no benefit, to destroy you is no loss." These ideologies didn't produce particularly healthy societies and were a fiasco from a biological standpoint. On the other hand, their antireligious attitude was part of a larger picture. All three countries had gone through an overthrow of the existing order, which may have required them to curb the power of established religion. I would not necessarily blame their atrocities on atheism per se, therefore. In the same way, killing in the name of God, such as during the Crusades or by the Spanish Conquistadores, was often a front for political or colonial ambitions. Columbus's lust for gold matched his love of God. To single out religion as the cause is therefore problematic. The bottom line is that humans are capable of unbelievable cruelty, whether in the name of God or while denying his existence.

Perhaps the question can be answered on a smaller scale, as in a study of the longevity of nineteenth-century communities in the United States. Communities based on a secular ideology, such as collectivism, disintegrated much faster than those based on religious principles. For every year that communities lasted, religious ones were four times as likely to survive than their secular counterparts.[21] Sharing a religion dramatically raises trust. We know the huge bonding effect of coordinated practices, such as praying together and carrying out the same rituals. This relates to the primate principle that acting together improves relationships, ranging from monkeys' preferring human experimenters who imitate them to varsity rowers' gaining physical resistance (such as a higher pain threshold) from exercising as a team rather than on their own.[22] Joint action may stimulate endorphin release, as has also been suggested for other bonding mechanisms, such as joint laughter. These positive effects of synchronization help explain the cohesiveness of religions and their effect on social stability.

Durkheim dubbed the benefits derived from belonging to a religion its "secular utility." He was convinced that something as pervasive and ubiquitous as religion must serve a purpose—not a higher purpose, but a social one. The biologist David Sloan Wilson, who analyzed the data on early Christians, agrees in that he sees religion as an adaptation that permits human groups to function harmoniously: "Religions exist primarily for people to achieve together what they cannot achieve alone."[23]

Religious community building comes naturally to us. In fact, given how commonly religion is pitted against science, it is good to remember the tremendous advantage religion enjoys. Science is an artificial, contrived achievement, whereas religion comes as easily to us as walking or breathing. This has been pointed out by many authors, from the American primatologist Barbara King, who in *Evolving God* relates the drive to religion to our desire to belong, to the French anthropologist Pascal Boyer, who views religion as an intuitive capacity:

> *Scientific research and theorizing has appeared only in very few human societies. . . . The results of scientific research may be well-known, but the whole intellectual style that is required to achieve them is really difficult to acquire. By contrast, religious representa-*

> *tions have appeared in all human groups that we know, they are easily acquired, they are maintained effortlessly and they seem accessible to all members of a group, regardless of intelligence or training. As Robert McCauley points out, . . . religious representations are highly natural to human beings, while science is quite clearly unnatural. That is, the former goes with the grain of our evolved intuitions, while the latter requires that we suspend, or even contradict most of our common ways of thinking.*[24]

Contrast the ease with which children adopt religion with the long and laborious road young people travel to achieve a Ph.D. around the age of thirty. McCauley, a philosopher colleague of mine at Emory, told me that if he had to choose which of the two would survive if society collapsed, he'd put his money on religion rather than science: "Religion overwhelmingly depends upon what I'm calling natural cognition, thinking that is automatic, that is not conscious for the most part." McCauley contrasts this with science, which is "conscious, usually in the form of language. It's slow, it's deliberative."[25]

Imagine, we put a few dozen children on an island without adults. What would happen? William Golding thought he knew, giving us savagery and murder in *Lord of the Flies.* This may have been a great extrapolation from life at English boarding schools, but there is no shred of evidence that this is what children left to their own devices will do. When four- to five-year-old children are left alone in a room, they tend to negotiate with each other by means of moral terminology such as "That's not fair!" or "Why don't you give her some of your toys?"[26] No one knows what children would do if left alone for a much longer time, but they would definitely form a dominance hierarchy. Young animals, whether goslings or puppies, quickly battle it out to establish a pecking order, and children do the same. I remember the pale faces of psychology students steeped in academic egalitarianism, upset at seeing young children beat up on each other on the first day of preschool. We are a hierarchical primate, and, however much we try to camouflage it, it comes out early in life.

The children on the island might also enter the symbolic domain. They'd probably develop language in the same way that Nicaraguan deaf children, in the 1980s, began communicating in a simple sign language that outsiders couldn't follow. Many other aspects would develop as well, such as culture. The children would transmit habits and knowledge and show conformism in the tools they made or how they greeted each other. They'd also have property rights and tensions over ownership. Finally, they would undoubtedly develop religion. We don't know what kind, but they would believe in supernatural forces, perhaps personalized ones, like gods, and develop rituals to appease them and bend them to their will.

The one thing the children would never develop is science. By all accounts, science is only a few thousand years old, hence it appeared extremely late in human history. It is a true accomplishment, a critically important one, yet it would be naïve to put it on the same level as religion. The war between science and religion is, to put it in biblical terms, one between David and Goliath. Religion has always been with us and is unlikely to ever go away, since it is part of our social skin. Science is rather like a coat that we have recently bought. We always risk losing it or throwing it away. The antiscience forces in society require constant vigilance, given how fragile science is compared with religion. To contrast the two as if they are on equal footing and in competition is a curious misrepresentation explainable only by reducing science and religion to sources of knowledge about the same phenomena. Only then could anyone argue that if one of the two is right, the other must be wrong.

When it comes down to knowledge of the physical world, the choice is obvious. I can't fathom why in this day and age, with everyone walking around with laptops and traveling through the air, we still need to defend science. Consider how far biomedical science has come, and how much longer we live as a result. Isn't it obvious that science is a superior way of finding out how things work, where humans come from, or how the universe arose? I am among scientists every day, and there is nothing more addictive than the thrill of discovery. True, plenty of mysteries are left, but science offers the only realistic hope of solving them. Those who present religion as a source of this kind of knowledge, and stick to age-old stories despite the avalanche of new information, deserve all the scorn they invite. But I also consider this particular collision between science and religion a mere sideshow. Religion is much more than belief. The question is not so much whether religion is true or false, but how it shapes our lives, and what might possibly take its place if we were to get rid of it the way an Aztec priest rips the beating heart out of a virgin. What could fill the gaping hole and take over the removed organ's functions?

I once saw an off-Broadway play entitled *Freud's Last Session*, in which the coughing and cigar-smoking psychoanalyst confronted C. S. Lewis, who had become a devout Christian, challenging the younger man's convictions. It was a dazzling display of skepticism, which made my subsequent reading of the real Freud a bit sobering. Even though Freud dismissed religion in no uncertain terms as a human creation and a mere "illusion," he was unwilling to recommend its abandonment. Freud waited until the end of *The Future of an Illusion* to let us feel the coldness of his feet:

> *If you wish to expel religion from our European civilization you can only do it through another system of doctrines, and from the outset this would take over all the psychological characteristics of religion, the same sanctity, rigidity, and intolerance, the same prohibition of thought in self-defense.*[27]

Was not the entire communist experiment an attempt at a godless society, and did it not follow Freud's predictions to the letter? With its sing-alongs, marching, reciting of pledges, and waving in the air of Little Red Books, the movement deliberately mimicked religion. Dogmatism, rigidity, and unholy fervor were on full display and grew with the decades, until communism collapsed under its own weight and lack of success. Having witnessed the early stages of this experiment, Freud may have guessed its futility.

Another curious attempt at going godless took place in 1793, when the altar in the Notre Dame cathedral was replaced with a model mountain on top of which stood a temple dedicated to philosophy. Next to it burned the "Torch of Truth." The Cult of Reason abolished Sunday as the seventh day of rest (replacing it with the tenth day), secularized the names of all saint's days, and squashed any hope of an afterlife by printing "Death is an eternal sleep" above the gates of cemeteries. The cult had its own goddess, a classically dressed lady who was carried through the streets of Paris followed by a throng of palm-leaf-waving acolytes. The procession brought her to the "mountain" in the cathedral, where she was seated between the busts of Voltaire and Rousseau. The cult ended abruptly when its leaders were executed by Maximilien Robespierre. Robespierre then started his own Cult of the Supreme Being, with himself as high priest. Immortality of the soul was quickly reinstated, which was a good thing, given how many innocents Robespierre sent to the guillotine. This cult, too, enjoyed only a short life—as short as that of its high priest.

Freud put his finger on an eternal pendulum swing in Western thought, which for centuries has moved back and forth between the mockery of religion as irrational and an "opium of the people," as Karl Marx put it, followed by concern about what would happen if we were to erase it from our lives. The neo-atheists have by now rehashed all of the arguments amassed over the centuries against religion. Hitchens showed himself a true Marxist with his "religion poisons everything," while Harris took over the Parisian Torch of Truth by yearning for a "religion of reason," and Dawkins's "delusion" hardly improves upon

Freud's "illusion."[28] Inevitably, however, we are now entering the cold-feet phase of the cycle. Apart from the question whether we'd even be capable of the self-amputation that atheists call for, the deeper issue is how to fill the God-sized vacuum if we succeed. Alain de Botton is an atheist who grudgingly admires religion for its understanding of universal human needs and weaknesses, while Philip Kitcher urges atheists and agnostics to go beyond disbelief. Criticizing religion is facile, Kitcher says, but any thinking human being about to join the atheist movement would want to know what it is for rather than what it is against:

> *Each of us needs an account of ourselves and what is valuable, something towards which we can steer and by which we can live. . . . [S]ecular thought shies away from the traditional question, raised by the Greeks at the dawn of philosophy, of what makes human lives, finite though they are, significant and worthwhile. . . . No advocacy of disbelief, however eloquent, will work the secular revolution until these facts are acknowledged. The temporary eradication of superstition, unaccompanied by attention to the functions religion serves, creates a vacuum into which the crudest forms of literalist mythology can easily intrude themselves. . . .*[29]

A Watchful Eye

On a recent visit to Vancouver, the Canadian psychologist Ara Norenzayan gave me the title of his new book, which I promptly wrote down as "Big Dogs." I may be slightly dyslexic, or perhaps it was a Freudian slip, my mind being more with animals than people. Ara had said "Big Gods."

He studies the role of religion in everyday life. One experiment investigated how "priming" people with religious thoughts affected generosity. Priming is the planting of an unconscious bias, hereby letting subjects correct the grammar of a few sentences that included words like "God," "prophet," and "divine." They encountered these words without any further information and had no idea what the experiment was about. After this, each subject found ten one-dollar coins on a table with the instruction to take as many as they liked, knowing that what they left behind would go to the next person. The outcome was spectacular. Unprimed subjects left on average only $1.84 for the other, but those primed with thoughts of God and religion left $4.22. About two-thirds of the primed group left more than half the coins behind. Curiously, religiosity didn't seem to matter much. Asked about their religion, about half the subjects answered "none," yet many of those performed like the rest.[30]

How to explain this effect? The thinking is that in large-scale societies, like ours, there is a need for supervision—imagined or real—to ensure a high level of cooperation. It is far too easy to get away with freeloading in an anonymous mass. The study participants probably conjured up a watching God in their heads, who approves of kindness and frowns upon cheating. "Watched people are nice people," explained Ara. This could also account for the "Sunday effect" on devout Christians, who on Sundays donate more money to good causes and watch less porno on the Internet.[31]

Having a supernatural supervisor may be a recent phenomenon, however, because during our prehistory we didn't really need one. In small groups, similar to primate groups, everyone knew everyone. Surrounded by kin, friends, and other community members, we had excellent reasons to follow the rules and get along with each other. We had personal reputations to uphold. It is only when our ancestors began to aggregate in ever larger societies, first with thousands of people, then with millions, that these face-to-face mechanisms fell apart. That is why Ara believes that with bigger groups came a need for bigger gods, who watch like hawks over everything we do. This nicely fits my own thinking that morality predates religion, certainly the dominant religions of today. We humans were plenty moral when we still roamed the savanna in small bands. Only when the scale of society began to grow and rules of reciprocity and reputation began to falter did a moralizing God become necessary.

In this view, it wasn't God who introduced us to morality; rather, it was the other way around. God was put into place to help us live the way we felt we ought to, confirming Voltaire's quip about our need to invent him. Think also of Socrates's question to Euthyphro whether an action is moral because the gods love it or whether the gods love moral actions. The whole purpose of God is to do the latter. We endowed him with the capacity to keep us on the same straight and narrow that we'd been following ever since we lived in small bands.

For those who despair at the implication that without religion the world might lack prosociality, there are a few rays of sunlight. First of all, the original experiment was incomplete. It primed for religious concepts only, not for any alternatives. This deficit was corrected in a second experiment in which subjects encountered good-citizen terminology, such as "civic," "jury," and "court" before being tested. Lo and behold, they became as altruistic as those primed with religious terms, leaving on average $4.44 on the table. This outcome offers hope for secular societies. If appeals to community values, the social contract, and law enforcement are as effective as religion at inducing generosity, the positive effects of religion may be replicable after all.

Second, a recent study compared the reasons why believers and nonbelievers assist others. It found nonbelievers to be more sensitive to the situation of others, basing their altruism on feelings of compassion. Believers, in contrast, seemed driven by a sense of obligation and how they ought to behave according to their religion. The behavioral outcome was the same, but the underlying motivations seemed different.[32] Clearly, there are many reasons for kindness, and religion is just one of them.

The secular model is currently being tried out in northern Europe, where it has progressed to the point that children naïvely ask why there are so many "plus signs" on large buildings called "churches," and where people have no idea anymore of the biblical origin of their expressions, from "washing your hands of the matter" to "a drop in the bucket." Civic institutions have taken over many of the functions originally fulfilled by the churches, such as care for the sick, poor, and old. Despite being largely agnostic or nonpracticing, the citizenry of these countries stands firmly behind this effort. It is a giant experiment, both economically and morally, that may tell us whether large nation-states can forge a well-functioning moral contract without religion. If one believes, as I do, that morality comes mostly from within, there is every reason to support this effort, but I also agree with Freud, Kitcher, and others that its success will require far more than God's death certificate.

Chapter 8

BOTTOM-UP MORALITY

> *'Tis to much purpose to go upon stilts, for, when upon stilts, we must yet walk with our legs; and, when seated upon the most elevated throne in the world, we are but seated upon our breech.*
>
> —Montaigne[1]

Entering Room 56 of the Museo Nacional del Prado, in Madrid, for the second time in my life was like walking into a shrine. After having read so much about *El Bosco*, and having recently toured the Jheronimus Bosch Art Center in my hometown with old friends, I was still pleasantly surprised by how colorful and gay the original *Garden* looked. The green and blue background, the red fruits, the flamboyant birds, and the mass of pale pink (and a handful of black) nudes created a delightful, festive atmosphere. All the more so since I had entered the room at the side where *The Triumph of Death* hung, a grim painting so brown and dull that one feels like dying right then and there. That was of course the artist's intention, since Brueghel was perfectly capable of producing vibrant colors.

Room 56 is a tall, well-lighted room in which *The Garden* is roped off, so that people can't get too close. Looking over the heads of a tourist group, I inspected the details of this one-hundred-paintings-in-one masterpiece, which has its own T-shirts, calendars, agendas, and mouse pads at the museum gift shop. I felt the way Antonio Damasio, the celebrated neuroscientist, must have felt making a pilgrimage in the opposite geographic direction for his book *Looking for Spinoza.* He went to see the Enlightenment philosopher's dwellings in The Hague and Rijnsburg, in the Netherlands.

Damasio wanted to learn more about Spinoza, who was a Dutchman of Portuguese descent. His Jewish parents had fled the Inquisition in their home country to avoid forced conversion. Born Portuguese himself, Damasio naturally related to Spinoza, whom he contrasted with that other philosopher Kant, by saying that whereas Kant wished to combat the perils of passion with reason, Spinoza considered passion the engine of his thinking. Damasio portrayed Spinoza as one of the most biologically oriented philosophers, while lamenting the disparagement of his approach because of his skepticism about the Abrahamic God. In much the same way, I visited the Prado to find Bosch's oeuvre in exile, feeling that the painter still has much to tell us about the question of moral origins and religion's role therein, even though Bosch, too, is rarely given his due. He deeply influenced surrealism, the movement celebrated four centuries after his death as new, exciting, and a sign of expanded consciousness. Bosch turned dreams into reality and illustrated the eternal foibles of humanity along the same lines as his contemporary Erasmus of Rotterdam did in writing. Erasmus's hugely popular *Praise of Folly* explained in erudite Latin how vain and naturally stupid the human race is. I cannot but sympathize with these early attempts to bring the crown of creation down to earth.

A fine example is the other Bosch triptych in Room 56, known as *The Hay Wain.* It shows a wagon with an enormous pile of hay passing through a crowd of people. On closer inspection, we see that the people are fighting over straws. In Middle Dutch, "hoy" (hay) signified vanity, nothingness, emptiness.[2] The painting shows people at each other's throats over mere hay, pulling knives and pummeling each other, while others are getting crushed under the wagon wheels. The clergy is very much part of the scramble, as a portly monk waits for nuns to fill his hay bag. Noblemen and the pope follow the wagon on horseback with all of their dignity intact to show that the haves don't need to mingle with the have-nots to get what they want. The wagon rolls on, luring everyone like a pied piper toward the right-hand panel, where hell awaits them. The triptych would make a perfect conversation piece for churches preaching the "prosperity gospel"—which promises financial blessings to the faithful—given Bosch's message that greed is ugly and pointless. This is obviously one of the oldest messages of Christian morality, as reflected in Jesus's statement that for a rich man to enter heaven will be harder than for a camel to go through the eye of a needle. The triptych exposes the corrupting influence of material wealth, presenting its pursuit as a life spent grasping at straws, literally.

Yet, it remains true that we barely understand Bosch, about whose life we know so little and about whose ideas and beliefs we know even less. We interpret him as best we can without any certainty that our deductions correspond with his intentions. As the great German art historian Erwin Panofsky concluded about Bosch, "We have bored a few holes through the door of the locked room, but somehow we do not seem to have discovered the key."[3]

Checking out *The Garden*'s left-hand panel, I was struck again by the astonishing suggestion that paradise may have been populated through a process other than creation. Not that Bosch was an evolutionist. Modern evolutionary ideas arose only in eighteenth-century France and England—before Darwin, but long after Bosch. Rather, the references in his triptych are to Aristotle's "spontaneous generation," according to which a rotting mixture of water, mud, and dung

Bosch depicted the spontaneous generation of strange and deformed creatures emerging from murky pools in this (slightly modified) fragment of *The Garden*. The scene moreover includes two instances of predation. Was the painter trying to be provocative by placing this evolutionary-looking tableau right at the feet of Adam and Eve?

is capable of producing live (but sometimes malformed) creatures. *The Garden* features two pools of primordial soup from which emerge feathered and flippered beasts, winged fish, a swimming unicorn, a three-headed bird, a seal with front legs, and a variety of amphibians. Naturally, water is an obsession for the Dutch, who live below sea level. Bosch often chose standing water to hint at evil, but here it produces life. I know of no other painting depicting such scenes, which cannot help but touch a biologist's heart. The high point is a duck-billed creature calmly reading a book, suggesting that the fruit of knowledge was actually available in the paradisiacal mud.

By putting emerging species next to Adam and Eve, the painter seems to suggest a connection between modest life forms and the creation of humanity. Often portrayed as a devout Christian, Bosch nevertheless sprinkled his works with the seeds of skepticism.

Lowly Beginnings

Animals crawling out of the mud recall our lowly beginnings. Everything started simple. This holds not only for our bodies—with hands derived from frontal fins and lungs from a swim bladder—but equally for our mind and behavior. The belief that morality somehow escapes this humble origin has been drilled into us by religion and embraced by philosophy. It is sharply at odds, however, with what modern science tells us about the primacy of intuitions and emotions. It is also at odds with what we know about other animals. Some say that animals are what they are, whereas our own species follows ideals, but this is easily proven wrong. Not because we don't have ideals, but because other species have them, too.

Why does a spider repair her web? It's because she has an ideal structure in mind, and as soon as her web deviates from it, she works hard to bring it back to its original shape. How does a "Mama Grizzly"[4] keep her young safe? Anybody moving between a sow and her cubs will discover that she has an ideal configuration in mind, which she doesn't like to be messed with. The animal world is full of repair and correction, from disturbed beaver dams and anthills to territorial defense and rank maintenance. Failing to obey the hierarchy, a subordinate monkey upsets the accepted order, and all hell breaks loose. Corrections are by definition normative: they reflect how animals feel things ought to be. Most pertinent for morality, which is also normative, social mammals strive for harmonious relationships. They are at pains to avoid conflict whenever they can. The gladiatorial view of nature is plainly wrong. In one field experiment, two fully grown male baboons refused to touch a peanut thrown between them, even though they both saw it land at their feet. Hans Kummer, the Swiss primatologist who worked all his life with wild hamadryas baboons, describes how two harem leaders, finding themselves in a fruit tree too small to feed both of their families, broke off their inevitable confrontation by literally *running away* from each other. They were followed by their respective females and offspring, leaving the fruit unpicked. Given the huge, slashing canine teeth of a baboon, few resources are worth a fight.[5] Chimp males face the same dilemma. From my office window, I often see several of them hang around a female with swollen genitals. Rather than competing, these males are trying to keep the peace. Frequently glancing at the female, they spend their day grooming each other. Only when everyone is sufficiently relaxed will one of them try to mate.

If fighting does break out, primates react the way the spider does to a torn web: they go into repair mode. Reconciliation is driven by the importance of social relationships. Studies on a great variety of species show that the closer two individuals are, and the more they do together, the more likely they are to make up after aggression.[6] Their behavior reflects awareness of the value of friendships and family bonds. This often requires them to overcome fear or suppress aggression. If it weren't for the need to bury the hatchet, it wouldn't make any sense for apes to kiss and embrace former opponents. The smart thing to do would be to stay away from them.

This brings me back to my bottom-up view of morality. The moral law is not imposed from above or derived from well-reasoned principles; rather, it arises from ingrained values that have been there since the beginning of time. The most fundamental one derives from the survival value of group life. The desire to belong, to get along, to love and be loved, prompts us to do everything in our power to stay on good terms with those on whom we depend. Other social primates share this value and rely on the same filter between emotion and action to reach a mutually agreeable modus vivendi. We see this filter at work when chimpanzee males suppress a brawl over a female, or when baboon males act as if they failed to notice a peanut. It all comes down to inhibitions.

Tara, the youngest female in our chimpanzee colony, has a naughty habit that drives the older ladies crazy. She sometimes finds a dead rat in the outdoor compound or digs one out of an abandoned hole. She then carries the corpse around by its tail, careful to keep it away from her body, and sneakily places it on the back or head of a sleeping group mate. Her victim jumps up as soon as she feels (or smells) the dead rat, loudly screaming and wildly shaking her body to get this ugly thing off of her. She may even rub the spot on her body with a fistful of grass, to make sure the smell is gone. Tara is quick to pick up her rat and go to her next target. What is most remarkable is the total lack of punishment. Her victims are highly upset, and Tara is at the bottom of the totem pole, yet she faces no consequences. She takes advantage of the extreme patience adults grant the young.

Emotional control comes in handy in life-or-death situations, such as the one described to me by Allan Schmidt, a lead caretaker at the Taronga Zoo, in Sydney. They have one of the world's nicest chimpanzee exhibits, in which two-year-old Sembe one day got herself entangled in a rope loop. Naturally, Sembe freaked out, and her screams quickly brought her mother, Shiba, to her aid. Shiba managed to disentangle the loop from Sembe and then took her daughter down to the ground to hold and comfort her. Once Sembe was reassured, Shiba climbed back up the rope and bit the offending loop, cutting

Grooming "bargaining" among three adult male chimpanzees near a sexually attractive female (foreground). Subordinate males have a better chance at an unchallenged mating after having groomed their superiors.

it so as to lessen future danger. Think about what it takes to free an infant from a deadly rope. The first impulse is no doubt to pull at the rope or the infant, which would surely make things worse. Instead, the mother provided the right assistance by loosening the loop, thus showing her understanding of its danger. The same understanding explains her subsequent safety measure.

We are mammals, a group of animals marked by sensitivity to each other's emotions. Even though I tend to favor primate examples, much of what I describe applies equally to other mammals. Take the work by the American zoologist Marc Bekoff, who analyzed videos of playing dogs, wolves, and coyotes. He concluded that canid play is subject to rules, builds trust, requires consideration of the other, and teaches the young how to behave. The highly stereotypical "play bow" (an animal crouches deep on her forelimbs while lifting her rear in the air) helps to set play apart from sex or conflict, with which it risks getting confused. Play ceases abruptly, though, as soon as one partner misbehaves or accidentally hurts the other. The transgressor "apologizes" by performing a new play bow, which may prompt the other to "forgive" the offense and continue to play. Role reversals make play even more exciting, such as when a dominant pack member rolls onto his back for a puppy, thus exposing his belly in an act of submission. This way, he lets the little one "win," something he'd never permit in real life. Bekoff, too, sees a relation with morality:

> *During social play, while individuals are having fun in a relatively safe environment, they learn ground rules that are acceptable to others—how hard they can bite, how roughly they can interact—and how to resolve conflicts. There is a premium on playing fairly and trusting others to do so as well. There are codes of social conduct that regulate what is permissible and what is not permissible, and the existence of these codes might have something to say about the evolution of morality.*[7]

For Bekoff, fair play refers to the way a dog ought to behave in order to be a good playmate. A big dog chasing a small one needs to pull its punches, and all dogs need to control their bite. These rules constitute what I have called one-on-one morality. But there is another way fairness comes into play, which concerns the division of resources. Even though all sorts of lofty principles regarding distributive justice have been formulated, the underlying emotions are, again, more basic than generally assumed. After all, even young children throw a fit if they get a smaller pizza slice than their sibling, shouting, "That's not fair!" They show first-order fairness, which is resentment at getting less than somebody else. Absent this emotion, why should anyone care about how things are divided?

The egalitarianism of hunter-gatherers suggests a long evolutionary history to our preoccupation with resource division. Hunters aren't even allowed to carve up their own kill, in order to prevent them from favoring family and friends. Anthropologists have played the Ultimatum Game across the globe and found humans everywhere to care about equity. The Ultimatum Game consists of two players dividing a sum between them. Only if both accept the split, however, will they get the money. Universally, our species favors an even split, probably because the party proposing the split realizes that he won't be able to get away with a skewed one. Brain scans of players facing unfair proposals reveal negative emotions, such as scorn and anger.[8]

The way humans play the Ultimatum Game is quite complex, because we not only show first-order fairness, which is protest at getting less, but anticipate this reaction in others and try to forestall it. We do so by actively promoting equity, thus reaching second-order fairness, which is a preference for fair outcomes in general. The critical role of conflict avoidance was already hinted at by Thomas Hobbes: "Every man is presumed to seek what is good for himselfe naturally, and what is just, onely for Peaces sake, and accidentally."[9] I agree with the political philosopher, except that I would never use the word "accidentally." A human tendency that is so pronounced and universal must be there for a reason.

How ancient this tendency is became clear when Sarah Brosnan and I discovered it in capuchin monkeys. This became an immensely popular experiment in which one monkey received cucumber slices while another received grapes for the same task. The monkeys had no trouble performing if both of them received identical rewards of whatever quality, but rejected unequal outcomes with such vehemence that there could be little doubt about their feelings. I often show their reactions to audiences, who almost fall out of their chairs laughing—which I interpret as a sign of surprised recognition.[10] Until then, they hadn't realized how closely their emotions resembled those of monkeys. The monkey receiving cucumber contentedly munches on her first slice, yet throws a tantrum after she notices that her companion is getting grapes. From then on, she ditches her measly cucumber slices and starts shaking the testing chamber with such agitation that it threatens to break apart. The underlying motivation is not so different from human street protest against unemployment or low wages. Occupy Wall Street is all about how some people roll in grapes while the rest of us live in cucumber land.

Refusing perfectly fine food because someone else is better off resembles human performance in the Ultimatum Game. Economists call this response "irrational," given that something is always better than nothing. No monkey, they say, should refuse food she'd otherwise eat, and no human should reject any offers, however small. Money is money. If these reactions are irrational, however, it is an irrationality that transcends species. To see it so vividly on display in a monkey helps us understand that our own sense of fairness, rather than being a product of our vaunted rationality, is rooted in basic emotions.

I should add, though, that our monkey experiment failed to show second-order fairness. We never saw the grape owner share her reward with the cucumber eater. This doesn't mean, however, that advanced fairness is uniquely human. We should also consider our closest relatives, the apes. First of all, apes regularly resolve conflicts over food that is not theirs. I once saw an adolescent female interrupt a quarrel between two juveniles over a leafy branch. She took the branch away from them, broke it into two, then handed each a part. Did she just want to stop the fight, or did she understand something about distribution? High-ranking males, too, often break up fights without taking any of the food that caused them. There's also the observation of Panbanisha, a bonobo, worried about being privileged. While being tested in a cognition laboratory, Panbanisha received large amounts of milk and raisins, but felt the envious eyes of her friends and family on her from a distance. After a while, she refused all rewards. Looking at the experimenter, she kept gesturing to the others until they, too, got some of the goodies. Only then did she finish hers. Apes have the ability to think ahead, and had Panbanisha publicly eaten her fill, there might have been negative repercussions when she rejoined the others later in the day.[11]

The most convincing evidence for second-order fairness, however, comes from Sarah's studies on chimpanzees. After we had carried out all of the additional tests that critics had asked for to prove that monkeys do indeed care about inequity, Sarah set up a large-scale chimpanzee project. We had discovered that sensitivity to fairness works only in combination with effort.[12] Just feeding primates unequally doesn't prompt any negative response. Rather, the food needs to serve as a payment for work. With this in mind, chimpanzees were rewarded with grapes and carrot pieces for a simple task, with grapes again being the favored food. As expected, carrot owners refused to perform or discarded their food if their partner received grapes. So far, the results confirmed the monkey work. But no one had anticipated that grape owners, too, would be upset. As Sarah wrote in her report, "We unexpectedly found that chimpanzees were more likely to refuse a high-value grape when the other chimpanzee got a lower-value carrot than when the other chimpanzee also received a grape."[13]

Fairness and justice are therefore best looked at as ancient capacities. They derive from the need to preserve harmony in the face of resource competition. We share both stages of fairness with the apes, and the first stage with monkeys and dogs. At the University of Vienna, Friederike Range found that dogs refuse to lift their paw for a "shake" with a human if they get nothing for it while a companion dog is rewarded.[14] We shouldn't be surprised to find this reaction in dogs, which derive from a long line of cooperative animals. Caring about what others get may seem petty, but in the long run it keeps one from getting duped. To call this response "irrational" misses the mark. If you and I often hunt together and you always take the best chunks of meat, I will need a different hunting buddy. It is probably no accident that all three animal species that dislike inequity—chimps, capuchins, and canids—are fond of meat and hunt in groups. Sensitivity to reward distribution helps ensure payoffs in line with effort, which is critical for sustained cooperation.

This brings me to next level of morality, the one where we leave the other primates behind. We care intensely about the group level and develop notions of right and wrong for everyone around, not just ourselves and our close relations. Not that this level is altogether absent in the apes—I have discussed it as "community concern"—but it requires greater powers of abstraction as well as the anticipation of what may happen if we let others get away with behavior that doesn't even directly affect us. We have the capacity to imagine its impact on the greater good. The underlying values are, again, not that complicated, because surely the functioning of a community is in the interest of all of its members, but it is harder to find parallels with other animals. We build reputations of honesty and trustworthiness, and disapprove of cheaters and noncooperators to the point of ostracizing them. Our goal is to keep everyone in line, putting collective interests before selfish ones. Morality serves to spread the benefits of group life around, and to keep a lid on exploitation by a powerful elite. Here, I am following the traditional view in biology, going back to Darwin, of morality as an in-group phenomenon. As Christopher Boehm has summarized it:

> *Our moral codes apply fully only within the group, be it a language group, a nonliterate population that shares the same piece of real estate or the same ethnic identity, or a nation. There seems to be a special, pejorative moral "discount" applied to cultural strangers—who often are not even considered to be fully human. . . .*[15]

But even if there can be little doubt that morality evolved for within-group reasons, without much consideration for humanity at large, this is not necessarily how it needs to be. Nowadays, we desperately try to move beyond moral parochialism and apply what we have learned about a dignified human life to the wider world, including strangers, even enemies. That enemies have rights, too, is a novel notion: the Geneva convention on prisoners of war stems only from 1929. The more we expand morality's reach, the more we need to rely on our intellect, because even though I believe that morality is firmly rooted in the emotions, biology has barely prepared us for rights and obligations on the scale of the modern world. We evolved as group animals, not global citizens. Nevertheless, we are well underway to reflect on these issues, such as universal human rights, and there is no reason to take the naturalized ethics advocated in this book as a prison from which we can't escape. It offers an account of how we got to where we are, but we humans have a long history of building new structures on top of old foundations.

The Bonobo and the Atheist

What would a bonobo tell an atheist? I have met the world's most language-proficient bonobo, Kanzi, who used to live in Atlanta together with his younger sister, Panbanisha. Even though Kanzi's understanding of spoken English is astonishing, and he is as smart a bonobo as I have ever seen, his utterances (on a computerized panel with symbols) are not at the level of academic debate. But let's pretend.

The bonobo would first of all urge the atheist to stop "sleeping furiously." There is no point getting all worked up about the absence of something, especially something as open to interpretation as God. True, if being a self-declared atheist carries a stigma, as it unfortunately does in this nation, frustration is understandable. Hatred breeds hatred, which is why some atheists rail against religion and talk as if its disappearance will be a huge relief. Never mind that religion is too deeply ingrained for it to ever be eliminated, and that historical attempts to do so by force have brought nothing but misery. Perhaps it can be done slowly and gently instead, but that would require us to appreciate and value our religious heritage at least to some degree, even if we regard it as outdated. Perhaps religion is like a ship that has carried us across the ocean, having allowed us to develop huge societies with a well-functioning morality. Now that we are spotting land, some of us are ready to disembark. But who says the land is as firm as it looks?

I am all for a reduced role of religion, with less emphasis on the almighty God and more on human potentials. This is nothing new, of course: it is the humanist agenda. Nowadays, humanism is often viewed as antireligious, but this is certainly not how it started.[16] Early humanism did criticize church theology, which it regarded as disconnected from practical life, but it was by-and-large compatible with Christian values. I should be careful here, though, because calling any values "religious" is problematic. It rather seems that universal human values have been appropriated by various religions, each supporting them with their own narratives and making them their own. It is only in the eighteenth century that humanism developed into an alternative to religion, gaining mass appeal by providing an ethical life stance based on reason instead of the supernatural. The fact remains, however, that humanism is nonreligious, not antireligious. Tolerance of religion, even if religion is not always tolerant in return, allows humanism to focus on what is most important, which is to build a better society based on natural human abilities. The outcome is the ongoing experiment in the West of an increasingly secular society. Like that of tectonic plates, the shifting is extremely gradual. Humanity cannot and will not change on a dime, and it's also not as if religion is an alien influence. It is very much our own creation, part of who we are, fully intertwined with our respective cultures. We had better get along with it and learn from it, even if our goal is ultimately to set out on a new course.

The bonobo would urge the atheist to take a similar long-term perspective. The good news is that the main ingredients of a moral society don't require religion, since they come from within. Despite its emphasis on reason, humanism considers our species a creature as much of passion as of intellect. This is where the bonobo has no trouble connecting. We have the emotions of a social animal, and not just any animal, but a mammal. Previous attempts at biological explanations of human behavior have suffered from too much emphasis on genes, and too many comparisons with social insects. Don't get me wrong, ants and bees are wonderful cooperators, and the study of them has greatly advanced our understanding of altruism. It is a triumph of evolutionary theory that its logic applies across such a vast array of species. Yet, insects possess none of the neural circuitry that mammals evolved for empathy and caring. Even if insect behavior resembles ours on the surface, it doesn't rely on the same processes. It's like comparing the chess play of computers and grandmasters: they may come up with the same moves, but get there in totally different ways.

The bonobo happily points out that he is no insect either. To compare ourselves with our fellow primates guards against reductive schemes according to which we are the slaves of our instincts. Those who think this way are quick to throw the word "mistake" around every time humans fail to follow the evolutionary party line. They find it easier to blame us than to revise their theories. The issue is that there are many layers between genes and behavior, from the encoding of proteins (which is what genes do) to neural processes and psychology. We are driven by inborn values and emotions, which guide rather than dictate behavior. They nudge us in a given direction, but leave plenty of leeway. As a result, we have the capacity to care for those unable to return the favor, adopt unrelated young, cooperate with strangers, and empathize with members of a different species. And we're not the only ones; the latest example was the assistance humpback whales gave to a mother gray whale defending her calf against a killer whale assault.[17] Mammals are affected by the distress of others, leading to levels of altruism far in excess of what gene-centric theories predict.

This is also why the bonobo disagrees with anyone who pits evolution against morality, such as the well-known American neurosurgeon Benjamin Carson, who has claimed, "Ultimately, if you accept the evolutionary theory, you dismiss ethics, you don't have to abide by a set of moral codes, you determine your own conscience based on your own desires."[18] The problem with such statements is that if humans everywhere develop a sense of right and wrong, one of our deepest desires must be to live in a moral world. Carson assumes that morality goes against our nature, and that our desires are all bad, whereas the whole point of this book is to argue the opposite. Thank God, if I may, we share with other primates a background as group animals, which makes us value social connections. Absent this background, religion could be preaching about virtue and vice until it became blue in the face, we'd never get its point. We are receptive only because of our evolved grasp of the value of relationships, the benefits of cooperation, the need for trust and honesty, and so on. Even our sense of fairness derives from this background.

Here the bonobo would side with the atheist and argue that whatever the role of religion in morality, it's a Johnny-come-lately role. Morality arose first, and modern religion latched onto it. Instead of giving us the moral law, the large religions were invented to bolster it. We are only just beginning to explore how religion does so by binding people together and enforcing good behavior. It is far from my intention to minimize this role, which was vital in the past and may remain so in the foreseeable future, but the wellspring of morality it is not.

Finally, the bonobo laughs at the intellectual torture of trying to separate "is" from "ought," which vexes any debate about moral evolution. It is widely accepted within philosophy that we cannot move from the way humans or animals are to moral ideals. The first is descriptive, they say, the second prescriptive. This is a serious consideration that is not easily resolved, but a good start would be to get our premises straight. If the thought is that animals are mere "wantons," lacking control over the impulses that nature has given them, we are on the wrong track. Like us, animals favor certain outcomes, and react with fear or violence to any deviation. Who says the bonobo can do whatever he wants? Even with regard to sex, where he faces fewer constraints than we do, he still needs a willing partner as well as the absence of dominant males. He faces many expectations about his conduct, which others won't hesitate to remind him of as soon as he scares an infant or tries to steal a female's food. Even if he lacks notions of right and wrong that transcend his personal situation, his values are not altogether different from those underlying human morality. He, too, strives to fit in, obeys social rules, empathizes with others, amends broken relationships, and objects to unfair arrangements. We may not wish to call it morality, but his behavior isn't free of prescriptions either.

With this observation, the bonobo concludes his advice to the atheist, whom he sees as a protester rather than an advocate. The big challenge is to move forward, beyond religion, and especially beyond top-down morality. Our best-known "moral laws" offer nice post hoc summaries of what we consider moral, but are limited in scope and full of holes. Morality has much more humble beginnings, which are recognizable in the behavior of other animals. Everything science has learned in the last few decades argues against the pessimistic view that morality is a thin veneer over a nasty human nature. On the contrary, our evolutionary background lends a massive helping hand without which we would never have gotten this far.

ACKNOWLEDGMENTS

It may seem a stretch to move from primate behavior to religion and humanism, but there is a logic to it. My interest in these issues started with studies on primate cooperation and conflict resolution, which led me to think about the evolution of empathy and, ultimately, human morality. My first book on this topic, *Good Natured* (1996), barely mentioned religion, but there are many people for whom morality and religion are inseparable, whereas others contest this connection. I felt it was time to add religious and nonreligious outlooks on life to the mix. They are essential to answering the question why our species so readily divides behavior into right and wrong.

There is also the inclusion of Hieronymus Bosch, who for me has always been a background presence. I named one of the Arnhem chimpanzees, Yeroen, after Bosch, whose first name in Dutch is Jeroen. Knowing my affection for the painter, students who worked with me at the time, in the 1970s, surprised me after my Ph.D. defense with a richly illustrated book on Bosch. Marianne Oertl, a German journalist and a painter herself, added to my interest by illuminating the connection between Bosch and my view of human nature. She saw him as an early humanist, which is also how I portray him in *The Bonobo and the Atheist.*

In 2009, the American anthropologist Sarah Hrdy and I received honorary doctorates from the University for Humanistics, in Utrecht, which further stimulated my exploration of the humanist angle in discussions with the philosopher Harry Kunneman and others. But, of course, the chief source of my approach to morality has always been my scientific work on the prosocial side of animal behavior. The decades of research that went into this book involved far too many co-workers, students, and funding sources to be mentioned by name. Let me thank my most recent collaborators and team members, who contributed to the findings reported, including the real-life stories that enliven my accounts: Kristin Bonnie, Sarah Brosnan, Sarah Calcutt, Matthew Campbell, Devyn Carter, Zanna Clay, Marietta Dindo, Tim Eppley, Pier Francesco Ferrari, Katie Hall, Victoria Horner, Kristi Leimgruber, Tara McKenney, Teresa Romero, Malini Suchak, Joshua Plotnik, Jennifer Pokorny, Amy Pollick, Darby Proctor, Diana Reiss, Taylor Rubin, Andy Whiten, and Yuko Hattori. I am grateful to the Yerkes National Primate Research Center of Emory University for the opportunity to conduct our studies, and feel grateful to the many monkeys and apes who have participated and become part of my life.

I have over the years interacted with many philosophers, who have sharpened my sense of how their discipline approaches morality. Philosophers have thought about it for several millennia, whereas biologists have only just begun. I thank all of them as well as other experts and friends for their advice and comments on parts of the manuscript: Isabel Behncke, Nathan Bupp, Patricia Churchland, Bettina Cothran, Peter Derkx, Ursula Goodenough, Orin Harman, Sarah Hrdy, Philip Kitcher, Harry Kunneman, Robert McCauley, Ara Norenzayan, Jared Rothstein, and Christopher Ryan. Thomas Vriens of the Jheronimus Bosch Art Center, in Den Bosch, fact-checked some of the sections on the painter, even though I remain solely responsible for the interpretations.

I thank my agent Michelle Tessler for her continued support, and my editor at Norton, Angela von der Lippe, for critical reading of the manuscript. As always, however, the commentator-in-chief has been Catherine, my wife, who eagerly reads my daily production and helps improve the text with her honest opinions. Even better, she pampers me and makes me a happy man.

NOTES

CHAPTER 1: **EARTHLY DELIGHTS**

1 Friedrich Nietzsche (1889), p. 5.

2 Also known as s'Hertogenbosch, the city is a twelfth-century provincial capital in the south of the Netherlands. At the time of Bosch's life (from circa 1450 until 1516), it was the second-largest city of the country, after Utrecht.

3 In 2007, Al Sharpton and Christopher Hitchens debated religion at the New York Public Library. See www.fora.tv.

4 The blog was posted on "The Stone," on 17 October 2010, available at http://opinionator.blogs.nytimes.com/2010/10/17/morals-without-god/.

5 Marc Kaufman, "Dalai Lama gives talk on science," *Washington Post*, 13 November 2005.

6 Advocating attention to chimpanzees rather than to bonobos, Melvin Konner (2002), p. 199, wrote, "And in any case, chimps have done far better than bonobos, which are very close to extinction."

7 "Hominin" is the new label for humans and their bipedal ancestors; previously it was "hominid."

8 Jürgen Habermas (2001) in his acceptance speech as recipient of the Peace Prize of the German Book Trade. The German text uses the word "Schuld," which means both blameworthy and being guilty: "Als sich Sünde in Schuld verwandelte, ging etwas verloren." Translated at www.csudh.edu/dearhabermas/habermas11.htm.

9 John Gray (2011), p. 235.
10 Allain de Botton (2012), p. 11.
11 Karl Pearson (1914), p. 91.
12 Sam Harris (2010) gave his book the subtitle *How Science Can Determine Human Values*.
13 PLoS Medicine Editors (2011).

CHAPTER 2: **GOODNESS EXPLAINED**

1 Charles Darwin (1871), p. 72.
2 Catherine Crockford et al. (2012).
3 J. B. S. Haldane, quoted in Oren Harman (2009), p. 158.
4 John Maynard Smith, quoted in Harman (2009), p. 167.
5 On Ronald Fisher and John von Neumann, see Harman (2009), p. 110.
6 From Frans de Waal (1996), p. 25.
7 George Price in a 1971 letter to Henry Morris, the father of modern creationism, quoted in Harman (2009), p. 248.
8 Robert Trivers in an interview with Carole Jahme in the *Guardian*, 7 October 2011.
9 Ernst Mayr (1997), p. 250, wrote, "Huxley, who believed in final causes, rejected natural selection and did not represent genuine Darwinian thought in any way. . . . It is unfortunate, considering how confused Huxley was, that his essay [on ethics] is often referred to even today as if it were authoritative."
10 For the statement by Joseph Hooker and other details of the Wilberforce-Huxley debate, see Ronald Numbers (2009), p. 155.
11 Letter of T. H. Huxley to Charles Darwin, 23 November 1859, in Leonard Huxley (1901), p. 189.
12 Letter of T. H. Huxley to Frederick Dyster, 10 October 1854, in Leonard Huxley (1901), p. 122.
13 From Leonard Huxley (1916), p. 322.
14 From T. H. Huxley (1894), p. 81.
15 Adrian Desmond (1994), p. 599.
16 Michael Ghiselin (1974), p. 247.
17 Robert Wright (1994), p. 344.
18 Defending his evaluation, George Williams (1988), p. 180, wrote, "I would concede that moral indifference might aptly characterize the physical universe. For the biological world a stronger term is needed."

19 An interview by Frans Roes (1997), p. 3, quotes Richard Dawkins: "What I am saying, along with many other people, among them T. H. Huxley, is that in our political and social life we are entitled to throw out Darwinism, to say we don't want to live in a Darwinian world."
20 Francis Collins (2006), p. 218.
21 Richard Dawkins on *Real Time with Bill Maher*, 11 April 2008.
22 From a letter by Charles Darwin to T. H. Huxley, 27 March 1882, in Desmond (1994), p. 519.
23 Charles Darwin (1871), p. 98.
24 Robert Boyd and Peter Richerson (2005) call this the "big mistake hypothesis."
25 Jessica Flack et al. (2005). De Waal (1992) provides data on impartial arbitration.
26 Roger Fouts and Stephen Mills (1997).
27 Jill Pruetz (2011).
28 Christophe Boesch et al. (2010).
29 Friedrich Nietzsche (1887), p. 51.
30 Marcus Aurelius (2002): "No one tires of being helped, and acts that are consistent with nature, like helping others, are their own reward. How then can you grow tired of helping others when by doing so you help yourself?"
31 Lara Aknin et al. (2012).

CHAPTER 3: **BONOBOS IN THE FAMILY TREE**

1 Common paraphrasing of a remark Abraham Lincoln made after he had spoken sympathetically of Southern rebels. A woman complained that it would be better for him to destroy than befriend his enemies, to which Lincoln replied (Ury, 1993, p. 146), "Why, madam, do I not destroy my enemies when I make them my friends?"
2 Jean-Baptiste Lamarck (1809), p. 170.
3 Georges Cuvier's "Elegy of Lamarck" was read to the French Académie des Sciences, in Paris, on 26 November 1832.
4 *The Colbert Report*, 30 January 2008.
5 Frans de Waal (1997a), p. 84.
6 Richard Wrangham and Dale Peterson (1996), p. 204.
7 Frans de Waal (1989), p. 215.
8 Frans de Waal (1997a), p. 81.
9 Watch Isabel Behncke Izquierdo at www.ted.com/talks/.

10 Robert Ardrey (1961) wrote, "We were born of risen apes, not fallen angels, and the apes were armed killers besides."
11 Gottfried Hohmann, quoted in Parker (2007).
12 Susan Block, "Bonobo Bashing in the New Yorker," *Counterpunch*, 25 July 2007, available at http://www.counterpunch.org/2007/07/25/bonobo-bashing-in-the-new-yorker/.
13 Takeshi Furuichi in Parker (2007).
14 Glenn Shepard (2011).
15 Laurinda Dixon (1981).
16 Hieronymus is the Latin form of Jerome. The Dutch commonly use "Jeroen" as the first name of the painter, whereas he himself signed his work with "Jheronimus Bosch."
17 Desiderius Erasmus (1519), p. 66.
18 Takeshi Furuichi (2011), p. 136.
19 Gottfried Hohmann and Barbara Fruth (2011), p. 72.
20 Robert Yerkes (1925), p. 246: "If I were to tell of his [the bonobo's] altruistic and obviously sympathetic behavior towards Panzee [a sick chimpanzee] I should be suspected of idealizing an ape."
21 Spindle cells are also known as Von Economo Neurons, or VEN cells. See John Allman et al. (2002).
22 James Rilling et al. (2011), p. 369.
23 Kay Prüfer et al. (2012). For more on "bipolar apes," see de Waal (2005).
24 Harold Coolidge (1933), p. 56.

CHAPTER 4: **IS GOD DEAD OR JUST IN A COMA?**

1 Jonathan Swift (1667–1745) quoted in Maturin Murray Ballou (1872), p. 433.
2 The label "new atheists," whom I shall call neo-atheists, applies mostly to four best-selling critics of religion and their followers: Sam Harris, Dan Dennett, Richard Dawkins, and Christopher Hitchens.
3 A. C. Grayling in an interview in the *Guardian*, 3 April 2011.
4 Born in 1525 in the same region, Pieter Brueghel the Elder was greatly influenced by Bosch. He was less moralizing, however, and more interested in daily life. He lived and worked mostly in Antwerp and Brussels.
5 *The O'Reilly Factor*, 4 January 2011, at www.youtube.com/watch?v=2BCipg71LbI.
6 Christopher Hitchens (2007) prided himself, before his atheist turn, on

having been Anglican, educated as a Methodist, married into Greek Orthodoxy, a follower of Sai Baba, and married by a rabbi.

7 Christopher Hitchens (2007).

8 Sam Harris (2010), p. 74: "the especially low-hanging fruit of conservative Islam."

9 John Draper's *History of the Conflict between Religion and Science* (1874) and Andrew White's *A History of Warfare of Science with Theology in Christendom* (1896).

10 Olaf Blanke and Shahar Arzy (2005), p. 17.

11 Frans de Waal et al. (1996).

12 Observing that neo-atheists are "sure that the world would be a better place if religion were hastened to extinction," Dan Dennett (2006) adds, "I am still agnostic about that. I don't know what could be put in religion's place—or what would arise unbidden—so I am still eager to explore the prospect of reforming religion."

13 Joseph Smith (1938).

14 Bernard Barber (1961), p. 596.

15 Jerry A. Coyne, "Science and religion aren't friends," *USA Today*, 11 October 2010.

16 Michael Gazzaniga interviewed in *Annals of the New York Academy of Sciences* (The Year in Cognitive Neuroscience) 1224 (2011), p. 8.

17 Albert Einstein: "What I dislike in this kind of argumentation [by Niels Bohr and others] is the basic positivistic attitude, which from my point of view is untenable. . . . 'Being' is always something which is mentally constructed by us, that is something which we freely posit." Cited in Michael Dickson (1999), an article on the entanglement of theory and observation.

18 Matt Ridley (2001) describes the first display of apes at the London Zoo.

19 In a Gallup poll in the United States in May 2011, 30 percent responded that the Bible is the actual word of God, 49 percent that it is an inspired text, and 17 percent that it is a book of fables and legends.

20 I reviewed this book about rape in the *New York Times Book Review*, 2 April 2000.

21 In 2009, the Pew Research Center asked Americans whether they agreed "that evolution is the best explanation for the origin of human life on earth." Buddhists (81 percent) and Hindus (80 percent) agreed the most, medium were Catholics (58 percent) and mainline Protestants (51 percent), and lowest were evangelical Protestants (24 percent) and Mormons (22 percent).

22 From a discussion between Matthieu Ricard and his father, Jean-François

Revel, a prominent French philosopher, in which Ricard says that science makes "une contribution majeure à des besoins mineurs" (Revel and Ricard, 1997). Ricard recently added that he now feels differently, that science has greatly improved the human condition.

23 Leo Tolstoy (1882) sought answers to questions such as "What is the meaning of my life?" "What will come of my life?" "Why does everything exist that exists, and why do I exist?"

24 Albert Einstein wrote in a letter of 25 October 1929 (Jammer, 1999, p. 51), "We followers of Spinoza see our God in the wonderful order and lawfulness of all that exists and in its soul as it reveals itself in man and animal. It is a different question whether belief in a personal God should be contested. . . . I myself would never engage in such a task. For such a belief seems to me preferable to the lack of any transcendental outlook of life, and I wonder whether one can ever successfully render to the majority of mankind a more sublime means in order to satisfy its metaphysical needs."

25 Interview in 2011 with Paul Kurtz: www.superscholar.org/interviews/paul-kurtz/.

26 Charles Renouvier (1859), p. 390: "à proprement parler, il n'y a pas de certitude; il y a seulement des hommes certains."

27 Ursula Goodenough (1999).

28 John Steinbeck (1951), p. 178.

CHAPTER 5: **THE PARABLE OF THE GOOD SIMIAN**

1 Michel de Montaigne (1877), vol. 1, p. 94.

2 Keith Jensen et al. (2006), p. 1013: "Chimpanzees make their choices based solely on personal gain, with no regard for the outcomes of a conspecific."

3 Ernst Fehr and Urs Fischbacher (2003).

4 Joan Silk et al. (2005), p. 1359: "The absence of other-regarding preferences in chimpanzees may indicate that such preferences are a derived property of the human species tied to [other] sophisticated capacities."

5 Victoria Horner et al. (2011a).

6 John Skoyles (2011); see reply by Victoria Horner et al. (2011b).

7 Revenge tactics were statistically confirmed by Frans de Waal (1992).

8 David Freedberg and Vittorio Gallese (2007), p. 197.

9 The first neurophysiological demonstration of human mirror neurons by Roy Mukamel et al. (2010).

10 B. F. Skinner (1953), p. 160.
11 Temple Grandin and Catherine Johnson (2004), p. 11.
12 Ivan Norscia and Elisabetta Palagi (2011).
13 Frans de Waal (2009), p. 61.
14 In a letter to Joshua Speed of 24 August 1855, Lincoln wrote, "[It is] a thing which has, and continually exercises, the power of making me miserable." Available at http://showcase.netins.net/web/creative/lincoln/speeches/speed.htm.
15 Grit Hein et al. (2010).
16 Dale Langford et al. (2006).
17 Tony Buchanan et al. (2011).
18 Inbal Ben-Ami Bartal et al. (2011), p. 1429.
19 John Darley and Daniel Batson (1973).
20 Teresa Romero et al. (2010).
21 Shinya Yamamoto et al. (2012).
22 Jill Pruetz and Stacy Lindshield (2011).

CHAPTER 6: **TEN COMMANDMENTS TOO MANY**

1 Immanuel Kant (1788), *Critique of Practical Reason*, available at www.gutenberg.org/cache/epub/5683/pg5683.html.
2 Edward Westermarck (1912), p. 19.
3 According to French media, Tristane Banon compared DSK to a "chimpanzé en rut."
4 Philip Kitcher (2006), p. 136, borrowed the term "wanton" from Harry Frankfurt (1971), who discussed it in the context of free will.
5 Christophe Boesch (2010).
6 Klaus Scherer (1994), p. 127.
7 For relative cortex size, see Katerina Semendeferi et al. (2002) and Suzana Herculano-Houzel (2009).
8 For delayed gratification in apes, see Theodore Evans and Michael Beran (2007).
9 Jesse Prinz (2006), p. 37.
10 Konrad Lorenz (1960).
11 Chris Coe and Leonard Rosenblum (1984) describe the experiment in which subordinate males reacted "in a manner that acknowledges a perceived violation of the social code."

12 From Frans de Waal (1982), p. 92.
13 Jessica Flack et al. (2004).
14 See Kevin Langergraber et al. (2007), Joan Silk et al. (2009), and Carl Zimmer, "Friends with benefits," *Time*, 20 February 2012.
15 David Hume (1739) notes how often authors move from descriptions of how things are to statements of how things ought to be, adding, p. 335, "This change is imperceptible; but is however, of the last consequence. For as this ought, or ought not, expresses some new relation or affirmation, 'tis necessary that it should be observed and explained; and at the same time that a reason should be given; for what seems altogether inconceivable, how this new relation can be a deduction from others, which are entirely different from it."
16 Patricia Churchland (2011).
17 Eric de Bruyn (2010).
18 Wilhelm Fränger (1951).
19 Henry Miller (1957), p. 29.
20 Philip Kitcher (2011), p. 207.
21 Favorable parole board rulings by Israeli judges were close to 0 percent before their lunch break, but about 65 percent thereafter (Danziger et al., 2011).
22 Blaise Pascal (1669): "Le cœur a ses raisons, que la raison ne connaît point," from *Pascal's Pensées,* Gutenberg ebook, www.gutenberg.org.
23 Claudia von Rohr et al. (2012).
24 Francys Subiaul et al. (2008).
25 Victoria Horner et al. (2010).
26 Edward Westermarck (1917), p. 238.
27 Edward Westermarck (1912), p. 38.
28 Christopher Boehm (2012).
29 Richard Lee (1969).
30 Michael Alvard (2004) and Joseph Henrich et al. (2001).
31 Shortened from Milton Diamond (1990), p. 423.
32 On human sexual variety, including comparisons with bonobos, see Sarah Hrdy (2009), Christopher Ryan and Cacilda Jethá (2010), Robert Walker et al. (2010), and Frans de Waal (2005).
33 Georgia representative Lynn Westmoreland on *The Colbert Report*, 14 June 2006.
34 Christopher Hitchens (2007), p. 99.
35 Sam Harris (2010).

36 The British philosopher Simon Blackburn made this "fools paradise" argument in a public rebuttal of Harris, in 2010: www.youtube.com/watch?v=W8vYq6Xm2To&feature=related.
37 Recognition of kinship by plants by Susan Dudley and Amanda File (2007).
38 Debate during the Tanner Lectures at Princeton University, in 2004. See Stephen Macedo and Josiah Ober (2006).
39 Michael Specter, "The dangerous philosopher," *The New Yorker*, 6 September 1999.
40 The American philosopher Mark Johnson (1993), p. 5, wrote, "It is morally irresponsible to think and act as though we possess a universal, disembodied reason that generates absolute rules, decision-making procedures, and universal or categorical laws by which we can tell right from wrong in any situation we encounter."
41 This debate is too complex to be covered in detail by a nonphilosopher, such as myself. Apart from the books mentioned by Kitcher and Churchland, I recommend Martha Nussbaum (2001) and Richard Joyce (2005).
42 Bonobo stories from the Milwaukee County Zoo, told by the ape caretaker Barbara Bell to Jo Sandin (2007) and myself.

CHAPTER 7: **THE GOD GAP**

1 Voltaire (1768), p. 402: "Si Dieu n'existait pas, il faudrait l'inventer."
2 Following chronic infection, Makali's finger had to be amputated by a veterinarian.
3 Geza Teleki (1973).
4 James Anderson et al. (2010), p. R351.
5 Nahoko Tokuyama et al. (2012).
6 Bert Haanstra's 1984 film documentary, *The Family of Chimps.*
7 Jeffrey Levin (1994) and William Strawbridge et al. (1997).
8 Jane Goodall (2005), p. 1304.
9 Victoria Horner and Andrew Whiten (2005).
10 Philip Dray (2005).
11 Richard Wrangham and Dale Peterson (1996) and Richard Byrne (1995).
12 Mathias Osvath (2009).
13 Mathias and Helena Osvath (2008).
14 Natacha Mendes et al. (2007).
15 Tetsuro Matsuzawa (2011), p. 304.
16 Ara Norenzayan and Ian Hansen (2006).

17 Carl Linfert (1972).
18 BBC, *The Forum*, 10 October 2010.
19 Roberta Smith, "Just when you thought it was safe," *New York Times*, 16 October 2007.
20 Daniel Everett (2005), p. 30. Andrew Nevins et al. (2009) express doubt about the absence of myths and beliefs in the Pirahã.
21 Richard Sosis and Eric Bressler (2003).
22 Emma Cohen et al. (2010).
23 David Sloan Wilson (2002), p. 159.
24 Pascal Boyer (2010), p. 85.
25 Michael Fitzgerald, "Why science is more fragile than faith," *Boston Globe*, 8 January 2012. See also Robert McCauley (2011).
26 William Arsenio and Melanie Killen (1996).
27 Sigmund Freud (1928), p. 89.
28 Sam Harris said, "There would be a religion of reason," in answer to the question what a world without God might look like. Gary Wolf, "The church of the non-believers," *Wired*, November 2006.
29 Philip Kitcher (2009).
30 Ara Norenzayan and Azim Shariff (2008), and Shariff and Norenzayan (2007).
31 Deepak Malhotra (2010) and Benjamin Edelman (2009).
32 Laura Saslow et al. (2012).

CHAPTER 8: **BOTTOM-UP MORALITY**

1 Michel de Montaigne (1877), vol. 3, p. 499.
2 A fifteenth-century Dutch expression about life is "Tis al hoy en stof" (It's all hay and dust).
3 Erwin Panofsky (1966), p. 357.
4 The expression "Mama Grizzly" was made popular in 2008 by the vice-presidential candidate Sarah Palin.
5 Hans Kummer (1995).
6 Frans de Waal (2000).
7 Marc Bekoff (2001), p. 85.
8 Joe Henrich et al. (2001) and Alan Sanfey et al. (2003).
9 Thomas Hobbes (1651), p. 36.
10 The monkey-fairness video concludes my 2012 TED talk, available at www.ted.com/talks.

11 Panbanisha's incident is related in Frans de Waal (1997a) by Sue Savage-Rumbaugh, who believes that her bonobos are happiest when everyone receives the same.
12 Megan van Wolkenten et al. (2007).
13 Sarah Brosnan et al. (2010).
14 Friederike Range et al. (2008).
15 Christopher Boehm (2012).
16 Peter Derkx (2011).
17 Candace Calloway Whiting, "Humpback whales intervene in orca attack on gray whale calf," *Digital Journal*, 8 May 2012, available at http://digitaljournal.com/article/324348.
18 Jonathan Gallagher, "Evolution? No: A conversation with Dr. Ben Carson," *Adventist Review*, 26 February 2004.

BIBLIOGRAPHY

Aknin, L. B., J. K. Hamlin, and E. W. Dunn. 2012. Giving leads to happiness in young children. *PLoS ONE* 7:e39211

Allman, J., A. Hakeem, and K. Watson. 2002. Two phylogenetic specializations in the human brain. *Neuroscientist* 8:335–46.

Alvard, M. 2004. The Ultimatum Game, fairness, and cooperation among big game hunters. In *Foundations of Human Sociality: Ethnography and Experiments from Fifteen Small-Scale Societies*, ed. J. Henrich et al., pp. 413–35. London: Oxford University Press.

Anderson, J. R., A. Gillies, and L. C. Lock. 2010. Pan thanatology. *Current Biology* 20:R349–R351.

Ardrey, R. 1961. *African Genesis: A Personal Investigation into the Animal Origins and Nature of Man.* New York: Simon & Schuster.

Arsenio, W. F., and M. Killen. 1996. Conflict-related emotions during peer disputes. *Early Education and Development* 7:43–57.

Bakewell, M. A., P. Shi, and J. Zhang. 2007. More genes underwent positive selection in chimpanzee evolution than in human evolution. *Proceedings of the National Academy of Sciences USA* 104:7489–94.

Ballou, M. M. 1872. *Treasury of Thought.* Boston: Osgood.

Barber, B. 1961. Resistance by scientists to scientific discovery. *Science* 134:596–602.

Bartal, I. B.-A., J. Decety, and P. Mason. 2011. Empathy and pro-social behavior in rats. *Science* 334:1427–30.

Bekoff, M. 2001. Social play behaviour cooperation, fairness, trust, and the evolution of morality. *Journal of Consciousness Studies* 8:81–90.

Belting, J. 2005. *Hieronymus Bosch: Garden of Earthly Delights.* Munich: Prestel.

Blanke, O., and S. Arzy. 2005. The out-of-body experience: Disturbed self-processing at the Temporo-Parietal Junction. *Neuroscientist* 11:16–24.

Boehm, C. 2012. *Moral Origins: The Evolution of Virtue, Altruism, and Shame.* New York: Basic Books.

Boesch, C. 2010. Patterns of chimpanzee's intergroup violence. In *Human Morality and Sociality: Evolutionary and Comparative Perspectives*, ed. H. Høgh-Olesen, pp. 132–59. Basingstoke, UK: Palgrave Macmillan.

Boesch, C., C. Bolé, N. Eckhardt, and H. Boesch. 2010. Altruism in forest chimpanzees: The case of adoption. *PLoS ONE* 5:e8901.

Bowles, S., and H. Gintis. 2003. The origins of human cooperation. In *The Genetic and Cultural Origins of Cooperation*, ed. P. Hammerstein, pp. 429–44. Cambridge, MA: MIT Press.

Boyd, R., and P. J. Richerson. 2005. Solving the puzzle of human cooperation. In *Evolution and Culture*, ed. S. Levinson, pp. 105–32. Cambridge, MA: MIT Press.

Boyer, P. 2010. *The Fracture of an Illusion: Science and the Dissolution of Religion.* Göttingen: Vandenhoeck & Ruprecht.

Brosnan, S. F., and F. B. M. de Waal. 2003. Monkeys reject unequal pay. *Nature* 425:297–99.

Brosnan, S. F., et al. 2010. Mechanisms underlying responses to inequitable outcomes in chimpanzees. *Animal Behaviour* 79:1229–37.

Brown, S. L., R. M. Nesse, A. D. Vinokur, and D. M. Smith. 2003. Providing social support may be more beneficial than receiving it: Results from a prospective study of mortality. *Psychological Science* 14:320–27.

Buchanan, T. W., S. L. Bagley, R. B. Stansfield, and S. D. Preston. 2011. The empathic, physiological resonance of stress. *Social Neuroscience* 7:191–201.

Byrne, R. 1995. *The Thinking Ape.* Oxford: Oxford University Press.

Churchland, P. S. 2011. *Braintrust: What Neuroscience Tells Us about Morality.* Princeton, NJ: Princeton University Press.

Coe, C. L., and L. A. Rosenblum. 1984. Male dominance in the bonnet macaque: A malleable relationship. In *Social Cohesion: Essays toward a Sociophysiological Perspective*, ed. P. R. Barchas and S. P. Mendoza, pp. 31–63. Westport, CT: Greenwood.

Cohen, E. E. A., R. Ejsmond-Frey, N. Knight, and R. Dunbar. 2010. Rowers' high: Behavioural synchrony is correlated with elevated pain thresholds. *Biology Letters* 6:106–8.

Collins, F. 2006. *The Language of God: A Scientist Presents Evidence for Belief.* New York: Free Press.

Coolidge, H. J. 1933. *Pan paniscus:* Pygmy chimpanzee from south of the Congo River. *American Journal of Physical Anthropology* 18:1–57.

Crockford, C., R. M. Wittig, R. Mundry, and K. Zuberbühler. 2012. Wild chimpanzees inform ignorant group members of danger. *Current Biology* 22:142–46.

Damasio, A. 2003. *Looking for Spinoza: Joy, Sorrow, and the Feeling Brain.* Orlando, FL: Harcourt.

Danziger, S., J. Leva, and L. Avnaim-Pesso. 2011. Extraneous factors in judicial decisions. *Proceedings of the National Academy of Sciences USA* 108:6889–92.

Darley, J. M., and C. D. Batson. 1973. From Jerusalem to Jericho: A study of situational and dispositional variables in helping behavior. *Journal of Personality and Social Psychology* 27:100–108.

Dart, R. A. 1953. The predatory transition from ape to man. *International Anthropological and Linguistic Review* 1:201–17.

Darwin, C. 1981 [orig. 1871]. *The Descent of Man, and Selection in Relation to Sex.* Princeton, NJ: Princeton University Press.

Dawkins, R. 1976. *The Selfish Gene.* Oxford: Oxford University Press.

———. 2006. *The God Delusion.* Boston: Houghton Mifflin.

de Botton, A. 2012. *Religion for Atheists: A Non-Believer's Guide to the Uses of Religion.* New York: Pantheon.

de Bruyn, E. 2010. Hieronymus Bosch's Garden of Delights triptych: The eroticism of its central panel and Middle Dutch. In *Jheronimus Bosch: His Sources*, pp. 94–106. s'Hertogenbosch: Jheronimus Bosch Art Center.

Dennett, D. 2006. Review of Richard Dawkins, *The God Delusion.* In *Free Inquiry.*

Derkx, P. 2011. *Humanisme, Zinvol Leven en Nooit meer "Ouder Worden."* Brussels: VUBPress.

Desmond, A. 1994. *Huxley: From Devil's Disciple to Evolution's High Priest.* Reading, MA: Perseus.

de Waal, F. B. M. 1989. *Peacemaking among Primates.* Cambridge, MA: Harvard University Press.

———. 1992. Coalitions as part of reciprocal relations in the Arnhem chimpanzee colony. In *Coalitions and Alliances in Humans and other Animals*, ed. A. Harcourt and F. B. M. de Waal, pp. 23–57. Oxford: Oxford University Press.

———. 1996. *Good Natured.* Cambridge, MA: Harvard University Press.

———. 1997a. *Bonobo: The Forgotten Ape.* Berkeley: University of California Press.

———. 1997b. The chimpanzee's service economy: Food for grooming. *Evolution and Human Behavior* 18:375–86.

———. 1999. Anthropomorphism and anthropodenial: Consistency in our thinking about humans and other animals. *Philosophical Topics* 27:255–80.

———. 2000. Survival of the rapist. Review of Thornhill and Palmer, *A Natural History of Rape*. In *New York Times Book Review*, 2 April, pp. 24–25.

———. 2005. *Our Inner Ape: A Leading Primatologist Explains Why We Are Who We Are.* New York: Riverhead.

———. 2007 [orig. 1982]. *Chimpanzee Politics: Power and Sex among Apes.* Baltimore, MD: Johns Hopkins University Press.

———. 2009. *The Age of Empathy: Nature's Lessons for a Kinder Society.* New York: Harmony.

de Waal, F. B. M., H. Uno, L. M. Luttrell, L. F. Meisner, and L. A. Jeannotte. 1996. Behavioral retardation in a macaque with autosomal trisomy and aging mother. *American Journal of Mental Retardation* 100:378–90.

Diamond, M. 1990. Selected cross-generational sexual behavior in traditional Hawai'i: A sexological ethnography. In *Pedophilia: Biosocial Dimensions*, ed. J. R. Feierman, pp. 422–44. New York: Springer.

Dickson, M. 1999. The light at the end of the tunneling: Observation and underdetermination. *Philosophy of Science* 66:47–58.

Dindo, M., B. Thierry, F. B. M. de Waal, and A. Whiten. 2010. Conditional copying fidelity in capuchin monkeys (*Cebus apella*). *Journal of Comparative Psychology* 124:29–37.

Dixon, L. S. 1981. Bosch's Garden of Delights: Remnants of a "fossil" science. *Art Bulletin* 63:96–113.

———. 2003. *Bosch.* London: Phaidon.

Douglas-Hamilton, I., S. Bhalla, G. Wittemyer, and F. Vollrath. 2006. Behavioural reactions of elephants towards a dying and deceased matriarch. *Applied Animal Behaviour Science* 100:87–102.

Dray, P. 2005. *Stealing God's Thunder: Benjamin Franklin's Lightning Rod and the Invention of America.* New York: Random House.

Dudley, S. A., and A. L. File. 2007. Kin recognition in an annual plant. *Biology Letters* 3:435–38.

Edelman, B. G. 2009. Red light states: Who buys online adult entertainment? *Journal of Economic Perspectives* 23:209–20.

Erasmus, D. 1519. *In Praise of Marriage.* In E. Rummel. 1996. *Erasmus on Women.* Toronto: University of Toronto Press.

Evans, T. A., and M. J. Beran. 2007. Chimpanzees use self-distraction to cope with impulsivity. *Biology Letters* 3:599–602.

Everett, D. L. 2005. Cultural constraints on grammar and cognition in Pirahã: Another look at the design features of human language. *Current Anthropology* 46:621–46.

Fehr, E., and U. Fischbacher. 2003. The nature of human altruism. *Nature* 425: 785–91.

Fehr, E., H. Bernhard, and B. Rockenbach. 2008. Egalitarianism in young children. *Nature* 454:1079–83.

Fessler, D. M. T. 2007. From appeasement to conformity: Evolutionary and cultural perspectives on shame, competition, and cooperation. In *The Self-Conscious Emotions: Theory and Research*, ed. J. L. Tracy, R. W. Robins, and J. P. Tangney, pp. 174–93. New York: Guilford.

Flack, J. C., L. A. Jeannotte, and F. B. M. de Waal. 2004. Play signaling and the perception of social rules by juvenile chimpanzees. *Journal of Comparative Psychology* 118:149–59.

Flack, J. C., D. C. Krakauer, and F. B. M. de Waal. 2005. Robustness mechanisms in primate societies: A perturbation study. *Proceedings of the Royal Society of London B* 272:1091–99.

Foerder, P., M. Galloway, T. Barthel, D. E. Moore, and D. Reiss. 2011. Insightful problem solving in an Asian elephant. *PLoS ONE* 6:e23251.

Fouts, R., and T. Mills. 1997. *Next of Kin.* New York: Morrow.

Fränger, W. 1976 [orig. 1951]. *The Millennium of Hieronymus Bosch: Outlines of a New Interpretation.* New York: Hacker Art Books.

Frankfurt, H. G. 1971. Freedom of the will and the concept of a person. *Journal of Philosophy* 68:5–20.

Freedberg, D., and V. Gallese. 2007. Motion, emotion and empathy in esthetic experience. *Trends in Cognitive Sciences* 5:197–203.

Freud, S. 2010 [orig. 1928]. *The Future of an Illusion.* Translated by W. D. Robson-Scott. Mansfield Centre, CT: Martino Publishing.

Furuichi, T. 2011. Female contributions to the peaceful nature of bonobo society. *Evolutionary Anthropology* 20:131–42.

Ghiselin, M. 1974. *The Economy of Nature and the Evolution of Sex.* Berkeley: University of California Press.

Goodall, J. 2005. Primate spirituality. In *The Encyclopedia of Religion and Nature*, ed. B. Taylor, pp. 1303–6. New York: Continuum.

Goodenough, U. 1999. The holes in Gould's semipermeable membrane between science and religion. *American Scientist*, May–June.

———. 2000. *The Sacred Depths of Nature*. New York: Oxford University Press.

Gould, S. J. 1997. Nonoverlapping Magisteria. *Natural History* 106(2):16–22

Grandin, T., and C. Johnson. 2004. *Animals in Translation: Using the Mysteries of Autism to Decode Animal Behavior*. New York: Scribner.

Gray, J. 2011. *The Immortalization Commission: Science and the Strange Quest to Cheat Death*. London: Allen Lane.

Haidt, J. 2001. The emotional dog and its rational tail: A social intuitionist approach to moral judgment. *Psychological Review* 108:814–34.

Hamlin, J. K., K. Wynn, and P. Bloom. 2007. Social evaluation by preverbal infants. *Nature* 450:557–59.

Hare, B., and S. Kwetuenda. 2010. Bonobos voluntarily share their own food with others. *Current Biology* 20:R230–R231.

Hare, B., et al. 2007. Tolerance allows bonobos to outperform chimpanzees on a cooperative task. *Current Biology* 17:1–5.

Harman, O. 2009. *The Price of Altruism*. New York: Norton.

Harris, S. 2006. *Letter to a Christian Nation*. New York: Knopf.

———. 2010. *The Moral Landscape: How Science Can Determine Human Values*. New York: Free Press.

Hein, G., G. Silani, K. Preuschoff, C. D. Batson, and T. Singer. 2010. Neural responses to ingroup and outgroup members' suffering predict individual differences in costly helping. *Neuron* 68:149–60.

Henrich, J., R. Boyd, S. Bowles, C. Camerer, H. Gintis, R. McElreath, and E. Fehr. 2001. In search of *Homo economicus:* Experiments in 15 small-scale societies. *American Economic Review* 91:73–79.

Herculano-Houzel, S. 2009. The human brain in numbers: A linearly scaled-up primate brain. *Frontiers in Human Neuroscience* 3:1–11.

Hitchens, C. 2007. *God Is Not Great: How Religion Poisons Everything*. New York: Hachette.

Hobaiter, C., and R. W. Byrne. 2010. Able-bodied wild chimpanzees imitate a motor procedure used by a disabled individual to overcome handicap. *PLoS ONE* 5:e11959.

Hobbes, T. 2004 [orig. 1651]. *De Cive*. Whitefish, MT: Kessinger.

Hohmann, G., and B. Fruth. 2011. Is blood thicker than water? In *Among African Apes*, ed. M. M. Robbins and C. Boesch, pp. 61–76. Berkeley: University of California Press.

Horner, V., and A. Whiten. 2005. Causal knowledge and imitation/emulation switching in chimpanzees (*Pan troglodytes*) and children (*Homo sapiens*). *Animal Cognition* 8:164–81.

Horner, V., D. J. Carter, M. Suchak, and F. B. M. de Waal. 2011a. Spontaneous prosocial choice by chimpanzees. *Proceedings of the Academy of Sciences USA* 108:13847–51.

———. 2011b. Reply to Skoyles: Misplaced assumptions of perfect human prosociality. *Proceedings of the Academy of Sciences USA* 108:E836.

Horner, V., D. Proctor, K. E. Bonnie, A. Whiten, and F. B. M. de Waal. 2010. Prestige affects cultural learning in chimpanzees. *PLoS-ONE* 5:e10625.

Hrdy, S. B. 2009. *Mothers and Others: The Evolutionary Origins of Mutual Understanding.* Cambridge, MA: Belknap Press of Harvard University Press.

Hume, D. 2008 [orig. 1739]. *A Treatise of Human Nature.* Sioux Falls, SD: NuVision.

Huxley, L., ed. 1901. *Life and Letters of Thomas Henry Huxley.* Vol. 1. New York: Appleton.

———. 1916. *Life and Letters of Thomas Henry Huxley.* Vol. 2. New York: Appleton.

Huxley, T. H. 1989 [orig. 1894]. *Evolution and Ethics.* Princeton, NJ: Princeton University Press.

Jammer, M. 1999. *Einstein and Religion.* Princeton, NJ: Princeton University Press.

Jensen, K., B. Hare, J. Call, and M. Tomasello. 2006. What's in it for me? Self-regard precludes altruism and spite in chimpanzees. *Proceedings of the Royal Society B* 273:1013–21.

Johnson, M. 1993. *Moral Imagination: Implications of Cognitive Science for Ethics.* Chicago: University of Chicago Press.

Joyce, R. 2005. *The Evolution of Morality.* Cambridge, MA: MIT Press.

Kano, T. 1992. *The Last Ape: Pygmy Chimpanzee Behavior and Ecology.* Stanford, CA: Stanford University Press.

King, B. J. 2007. *Evolving God: A Provocative View of the Origin of Religion.* New York: Doubleday.

Kitcher, P. 2006. Ethics and evolution: How to get here from there. In *Primates and Philosophers: How Morality Evolved*, ed. S. Macedo and J. Ober, pp. 120–39. Princeton, NJ: Princeton University Press.

———. 2009. Beyond Disbelief. In *50 Voices of Disbelief: Why We Are Atheists*, ed. R. Blackford and U. Schuklenk, pp. 87–96. Hoboken, NJ: Wiley-Blackwell.

———. 2011. *The Ethical Project.* Cambridge, MA: Harvard University Press.

Konner, M. 2002. Some obstacles to altruism. In *Altruistic Love: Science, Phi-*

losophy, and Religion in Dialogue, ed. S. G. Post et al., pp. 192–211. Oxford: Oxford University Press.

Kummer, H. 1995. *The Quest of the Sacred Baboon.* Princeton, NJ: Princeton University Press.

Lamarck, J. B. 1914 [orig. 1809]. *Zoological Philosophy.* Translated by Hugh Elliot. London: Macmillan.

Langergraber, K. E., J. C. Mitani, and L. Vigilant. 2007. The limited impact of kinship on cooperation in wild chimpanzees. *Proceedings of the Academy of Sciences USA* 104:7786–90.

Langford, D. J., et al. 2006. Social modulation of pain as evidence for empathy in mice. *Science* 312:1967–70.

Lee, R. B. 1969. Eating Christmas in the Kalahari. *Natural History* 78(12):14–22, 60–63.

Levin, J. S. 1994. Religion and health: Is there an association, is it valid, and is it causal? *Social Science and Medicine* 38:1475–82.

Linfert, C. 2003 [orig. 1972]. *Hieronymus Bosch.* New York: H. N. Abrams.

Lorenz, K. 1960. *So kam der Mensch auf den Hund.* Vienna: Borotha-Schoeler.

Macedo, S., and J. Ober, eds. 2006. *Primates and Philosophers: How Morality Evolved.* Princeton, NJ: Princeton University Press.

Malenky, R. K., and R. W. Wrangham. 1994. A quantitative comparison of terrestrial herbaceous food consumption by *Pan paniscus* in the Lomako Forest, Zaire, and *Pan troglodytes* in the Kibale Forest, Uganda. *American Journal of Primatology* 32:1–12.

Malhotra, D. 2010. (When) are religious people nicer? Religious salience and the "Sunday Effect" on pro-social behavior. *Judgment and Decision Making* 5:138–43.

Marcus Aurelius. 2002 [orig. 170–180 CE]. *The Emperor's Handbook: A New Translation of The Meditations.* New York: Scribner.

Matsuzawa, T. 2011. What is uniquely human? A view from comparative cognitive development in humans and chimpanzees. In *The Primate Mind*, ed. F. B. M. de Waal and P. F. Ferrari, pp. 288–305. Cambridge, MA: Harvard University Press.

Mayr, E. 1997. *This Is Biology: The Science of the Living World.* Cambridge, MA: Harvard University Press.

McCauley, R. N. 2011. *Why Religion Is Natural and Science Is Not.* New York: Oxford University Press.

Mendes, N., D. Hanus, and J. Call. 2007. Raising the level: Orangutans use water as a tool. *Biology Letters* 3:453–55.

Mercader, J., et al. 2007. 4,300-year-old chimpanzee sites and the origins of percussive stone technology. *Proceedings of the National Academy of Sciences USA* 104:3043–48.

Mercier, H., and D. Sperber. 2011. Why do humans reason? Arguments for an argumentative theory. *Behavioral and Brain Sciences* 34:57–111.

Midgley, M. 2010. *The Solitary Self: Darwin and the Selfish Gene.* Durham, UK: Acumen.

Miller, H. 1957. *Big Sur and the Oranges of Hieronymus Bosch.* New York: New Directions.

Montaigne, M. 1902 [orig. 1877]. *Essays of Montaigne.* Translated by Charles Cotton. Vols. 1–4. London: Reeves & Turner.

Mukamel, R., A. D. Ekstrom, J. Kaplan, M. Iacoboni, and I. Fried. 2010. Single neuron responses in humans during execution and observation of actions. *Current Biology* 20:750–56.

Muscarella, F., and M. R. Cunningham. 1996. The evolutionary significance and social perception of male pattern baldness and facial hair. *Ethology and Sociobiology* 17:99–117.

Nevins, A., D. Pesetsky, and C. Rodrigues. 2009. Pirahã exceptionality: A reassessment. *Language* 85:355–404.

Nietzsche, F. 2006 [orig. 1887]. *On the Genealogy of Morality and Other Writings.* Student Edition. Cambridge: Cambridge University Press.

———. 2007 [orig. 1889]. *Twilight of the Idols with the Antichrist and Ecce Homo.* Hertfordshire, UK: Wordsworth.

Norenzayan, A., and I. G. Hansen. 2006. Belief in supernatural agents in the face of death. *Personality and Social Psychology Bulletin* 32:174–87.

Norenzayan, A., and A. F. Shariff. 2008. The origin and evolution of religious prosociality. *Science* 322:58–62.

Norscia, I., and E. Palagi. 2011. Yawn contagion and empathy in *Homo sapiens. PLoS ONE* 6:e28472.

Nowak, M., and R. Highfield. 2011. *SuperCooperators: Altruism, Evolution, and Why We Need Each Other to Succeed.* New York: Free Press.

Numbers, R., ed. 2009. *Galileo Goes to Jail and Other Myths about Science and Religion.* Cambridge, MA: Harvard University Press

Nussbaum, M. C. 2001. *Upheavals of Thought: The Intelligence of Emotions.* Cambridge, UK: Cambridge University Press.

Osvath, M. 2009. Spontaneous planning for future stone throwing by a male chimpanzee. *Current Biology* 9:R190–R191.

Osvath, M., and H. Osvath. 2008. Chimpanzee (*Pan troglodytes*) and orangutan (*Pongo abelii*) forethought: Self-control and pre-experience in the face of future tool use. *Animal Cognition* 11:661–74.

Oxnard, C., P. J. Obendorf, and B. J. Kefford. 2010. Post-cranial skeletons of hypothyroid cretins show a similar anatomical mosaic as *Homo floresiensis*. *PLoS ONE* 5:e13018.

Panofsky, E. 1966. *Early Netherlandish Painting: Its Origins and Character.* Cambridge, MA: Harvard University Press.

Parker, I. 2007. Swingers. *The New Yorker*, July 30.

Parr, L. A., and F. B. M. de Waal. 1999. Visual kin recognition in chimpanzees. *Nature* 399:647–48.

Paukner, A., et al. 2009. Capuchin monkeys display affiliation toward humans who imitate them. *Science* 325:880–83.

Pearson, K. 1914. *The Life, Letters and Labours of Francis Galton.* London: Cambridge University Press.

Perelman, P., et al. 2011. A molecular phylogeny of living primates. *PLoS Genetics* 7:e1001342.

Perry, S. 2009. Conformism in the food processing techniques of white-faced capuchin monkeys (*Cebus capucinus*). *Animal Cognition* 12:705–16.

PLoS Medicine Editors. 2011. Medical complicity in torture at Guantánamo Bay: Evidence is the first step toward justice. *PLoS Medecine* 8:e1001028.

Plotnik, J. M., F. B. M. de Waal, and D. Reiss. 2006. Self-recognition in an Asian elephant. *Proceedings of the Academy of Sciences USA* 103:17053–57.

Plotnik, J. M., R. C. Lair, W. Suphachoksakun, and F. B. M. de Waal. 2011. Elephants know when they need a helping trunk in a cooperative task. *Proceedings of the Academy of Sciences USA* 108:5116–21.

Prinz, J. 2006. The emotional basis of moral judgments. *Philosophical Explorations* 9:29–43.

Prüfer, K., et al. 2012. The bonobo genome compared with the chimpanzee and human genomes. *Nature* 486:527–31.

Pruetz, J. D. 2011. Targeted helping by a wild adolescent male chimpanzee (*Pan troglodytes verus*): Evidence for empathy? *Journal of Ethology* 29:365–68.

Pruetz, J. D., and S. Lindshield. 2011. Plant-food and tool transfer among savanna chimpanzees at Fongoli, Senegal. *Primates* 53:133–45.

Range, F., L. Horn, Z. Viranyi, and L. Huber. 2008. The absence of reward induces inequity aversion in dogs. *Proceedings of the National Academy of Sciences USA* 106:340–45.

Renouvier, C. 1859. *Essais de critique générale. Deuxième essai. Traité de psychologie rationnelle d'après les principes du criticisme.* Paris: Ladrange.

Revel, J.-F., and M. Ricard. 1997. *Le moine et le philosophe.* Paris: Nil Editions.

Ribberink, E., and D. Houtman. 2010. Te ongelovig om atheïst te zijn: Over de-privatisering van ongeloof. *Religie & Samenleving* 5:209–26.

Ridley, M. 2001. Re-reading Darwin. *Prospect* 66:74–76.

Rilling, J. K., J. Scholz, T. M. Preuss, M. F. Glasser, B. K. Errangi, and T. E. Behrens. 2011. Differences between chimpanzees and bonobos in neural systems supporting social cognition. *Social Cognitive and Affective Neuroscience* 7:369–79.

Roes, F. 1997. An Interview of Richard Dawkins. *Human Ethology Bulletin* 12(1):1–3.

Romero, M. T., M. A. Castellanos, and F. B. M. de Waal. 2010. Consolation as possible expression of sympathetic concern among chimpanzees. *Proceedings of the National Academy of Sciences USA* 107:12110–15.

Ryan, C., and C. Jethá. 2010. *Sex at Dawn: The Prehistoric Origins of Modern Sexuality.* New York: Harper Collins.

Sandin, J. 2007. *Bonobos: Encounters in Empathy.* Milwaukee, WI: Zoological Society of Milwaukee.

Sanfey, A. G., J. K. Rilling, J. A. Aronson, L. E. Nystrom, and J. D. Cohen. 2003. The neural basis of economic decision-making in the ultimatum game. *Science* 300:1755–58.

Saslow, L. R., et al. 2012. My brother's keeper? Compassion predicts generosity more among less religious individuals. *Social Psychological and Personality Science.* doi:10.1177/1948550612444137.

Scherer, K. R. 1994. Emotion serves to decouple stimulus and response. In *The Nature of Emotion: Fundamental Questions*, ed. P. Ekman and R. J. Davidson, pp. 127–30. New York: Oxford University Press.

Schwartz, G. 2009. Visual arts: The humanist meets the exorcist. *ArtsFuse*, 21 February.

Semendeferi, K., A. Lu, N. Schenker, H. Damasio. 2002. Humans and great apes share a large frontal cortex. *Nature Neuroscience* 5:272–76.

Shariff, A. F., and A. Norenzayan. 2007. God is watching you: Priming God concepts increases prosocial behavior in an anonymous economic game. *Psychological Science* 18:803–9.

Shepard, G. H. 2011. The Mark and Olly Follies. *Anthropology News*, May, p. 18.

Silk, J. B., S. F. Brosnan, J. Vonk, J. Henrich, D. Povinelli, and A. S. Richardson. 2005. Chimpanzees are indifferent to the welfare of unrelated group members. *Nature* 437:1357–59.

Silk, J. B., et al. 2009. The benefits of social capital: Close social bonds among female baboons enhance offspring survival. *Proceedings of the Royal Society of London B* 276:3099–114.

Skinner, B. F. 1965 [orig. 1953]. *Science and Human Behavior.* New York: Free Press.

Skoyles, J. R. 2011. Chimpanzees make mean-spirited, not prosocial, choices. *Proceedings of the Academy of Sciences USA* 108:E835.

Smith, J. 1938. *Teachings of the Prophet Joseph Smith.* Edited by J. F. Smith. Salt Lake City: Deseret Book.

Smuts, B. B. 1985. *Sex and Friendship in Baboons.* New York: Aldine.

Sosis, R., and E. R. Bressler. 2003. Cooperation and commune longevity: A test of the costly signaling theory of religion. *Cross-Cultural Research* 37:211–39.

Strawbridge, W. J., R. D. Cohen, S. J. Shema, and G. A. Kaplan. 1997. Frequent attendance at religious services and mortality over 28 years. *American Journal of Public Health* 87:957–61.

Steinbeck, J. 1995 [orig. 1951]. *The Log from the Sea of Cortez.* New York: Penguin.

Subiaul, F., J. Vonk, J. Barth, and S. Okamoto-Barth. 2008. Chimpanzees learn the reputation of strangers by observation. *Animal Cognition* 11:611–23.

Tan, A. 2005. *Saving Fish from Drowning.* New York: Ballantine.

Tannen, D. 1990. *You Just Don't Understand: Women and Men in Conversation.* New York: Ballantine.

Teleki, G. 1973. Group response to the acidental death of a chimpanzee in Gombe National Park, Tanzania. *Folia primatologica* 20:81–94.

Thornhill, R., and C. T. Palmer. 2000. *The Natural History of Rape: Biological Bases of Sexual Coercion.* Cambridge, MA: MIT Press.

Tokuyama, N., et al. 2012. Bonobos apparently search for a lost member injured by a snare. *Primates* 53:215–19.

Tolstoy, L. 1961 [orig. 1882]. *A Confession.* Translated by A. Maude. London: Oxford University Press.

Trivers, R. L. 1971. The evolution of reciprocal altruism. *Quarterly Review of Biology* 46:35–57.

Turnbull, C. M. 1961. *The Forest People.* Garden City, NY: Natural History Press.

Ury, W. 1993. *Getting Past No: Negotiating with Difficult People.* New York: Bantam.

van Wolkenten, M., S. F. Brosnan, and F. B. M. de Waal. 2007. Inequity responses of monkeys modified by effort. *Proceedings of the National Academy of Sciences USA* 104:18854–59.

Voltaire. 1768. In *OEuvres Complètes de Voltaire.* Vol. 10. Paris: Garnier, 1877–85.

von Rohr, C. R., et al. 2012. Impartial third-party interventions in captive chimpanzees: A reflection of community concern. *PLoS ONE* 7:e32494.

Walker, R. S., M. V. Flinn, and K. R. Hill. 2010. Evolutionary history of partible paternity in lowland South America. *Proceedings of the National Academy of Sciences USA* 107:19195–200.

Warneken, F., B. Hare, A. P. Melis, D. Hanus, and M. Tomasello. 2007. Spontaneous altruism by chimpanzees and young children. *PLoS-Biology* 5:1414–20.

Westermarck, E. 1912 [orig. 1908]. *The Origin and Development of the Moral Ideas.* Vol. 1. 2nd ed. London: Macmillan.

———. 1917 [orig. 1908]. *The Origin and Development of the Moral Ideas.* Vol. 2. 2nd ed. London: Macmillan.

Whiten, A., V. Horner, and F. B. M. de Waal. 2005. Conformity to cultural norms of tool use in chimpanzees. *Nature* 437:737–40.

Wilkinson, A., N. Sebanz, I. Mandl, and L. Huber. 2011. No evidence of contagious yawning in the red-footed tortoise, *Geochelone carbonaria. Current Zoology* 57:477–84.

Williams, G. C. 1988. A sociobiological expansion of "Evolution and Ethics." In *Evolution and Ethics*, ed. J. Paradis and G. C. Williams, pp. 179–214. Princeton, NJ: Princeton University Press.

Wilson, D. S. 2002. *Darwin's Cathedral: Evolution, Religion, and the Nature of Society.* Chicago: University of Chicago Press.

Woods, V., and B. Hare. 2010. Bonobo but not chimpanzee infants use sociosexual contact with peers. *Primates* 52:111–16.

Wrangham, R. W., and D. Peterson. 1996. *Demonic Males: Apes and the Evolution of Human Aggression.* Boston: Houghton Mifflin.

Wright, R. 1994. *The Moral Animal: The New Science of Evolutionary Psychology.* New York: Pantheon.

Yamamoto, S., T. Humle, and M. Tanaka. 2012. Chimpanzees' flexible targeted helping based on an understanding of conspecifics' goals. *Proceedings of the National Academy of Sciences USA* 109:3588–92.

Yerkes, R. M. 1925. *Almost Human.* New York: Century.

INDEX

Page numbers in *italics* refer to illustrations.
Page numbers beginning with 245 refer to notes.

ABOUT THE AUTHOR

FRANS DE WAAL is a Dutch/American primatologist and biologist with a degree from the University of Utrecht. He has lived and worked in the United States since 1981. His first popular book, *Chimpanzee Politics*, compared the schmoozing and scheming of chimpanzees involved in power struggles with that of human politicians. Ever since, de Waal has drawn parallels between primate and human behavior. His books have reached a large audience in many different languages, making him one of the world's best-known primatologists.

With his discovery of reconciliation in primates, de Waal pioneered research on animal conflict resolution. He received the 1989 Los Angeles Times Book Award for *Peacemaking among Primates*. His scientific articles have been published in journals such as *Science*, *Nature*, *Scientific American*, and outlets specialized in animal behavior. De Waal is C. H. Candler Professor in the Psychology Department of Emory University and director of the Living Links Center at the Yerkes National Primate Research Center, in Atlanta. He is a member of both the U.S. National Academy of Sciences and the Royal Dutch Academy of Sciences. In 2007 he was selected by *Time* as one of The World's 100 Most Influential People Today, and in 2011 by *Discover* as among 47 (all-time) Great Minds of Science.

De Waal lives in Stone Mountain, Georgia, with his wife, Catherine.